Technological Innovation
and Multinational Corporations

Technological Innovation and Multinational Corporations

John Cantwell

Basil Blackwell

1989

Copyright © John Cantwell 1989

First published 1989

Basil Blackwell Ltd
108 Cowley Road, Oxford, OX4 1JF, UK

Basil Blackwell Inc.
3 Cambridge Center
Cambridge, Massachusetts 02142, USA

British Library Cataloguing in Publication Data
A CIP catalogue record for this book is available from the British Library.

Library of Congress Cataloging in Publication Data
Cantwell, John.
  Technological innovation and multinational corporations.
  Bibliography: p.
  Includes index.
  1. Technological innovations—Economic aspects.
2. Competition, International.
3. International business enterprises.
I. Title.
HC79.T4C36   1989   338.8′8   88–35079
ISBN 0–631–13847–1

Typeset in 11 on 13 pt. Sabon by
Vera-Reyes

Printed in Great Britain by Camelot Press Ltd., Southampton

Contents

List of Tables   ix
Preface   xiii

1 Introduction   1

   1.1 The Analysis of International Technological
        Competition   1
   1.2 The International Accumulation of
        Technology and Capital   6

   Notes   15

2 Historical Trends in International Patterns of
   Technological Innovation   16

   2.1 Introduction   16
   2.2 A Description of the Data   20
   2.3 The Statistical Methodology   25
   2.4 The Stability of Technological Advantage
        before 1914   31
   2.5 The Stability of Technological Advantage
        over the past Hundred Years   35
   2.6 The Stability of Technological Advantage since
        1963   37
   2.7 Conclusions and some Possible Extensions   45

   Notes   48

3 A Dynamic Model of the Post-war Growth
   of International Economic Activity in Europe
   and the USA   49

   3.1 Introduction   49

3.2 Innovation as a Dynamic Element in the
International Economic Activity of Firms   52
3.3 The Product Cycle Model (PCM) Revisited   54
3.4 The Explanation of International Production by
Innovative Manufacturing Firms before 1914   60
3.5 An Alternative Model of Technological
Competition between US and European Firms   61
3.6 The Application of the Model   68

Notes   71

4 The Evolution of Technological Competition between
US and European Firms   73

4.1 Introduction   73
4.2 European Catching Up, and the Relative
Growth of European and US Exports of
Manufactures, 1955–1975   75
4.3 The Role of US International Production in
European Catching Up, and the Response of
Indigenous European Firms   77
4.4 The Role of Licensing in the European Revival 88
4.5 Conclusions   91

Notes   92

5 The European and Japanese Response to the
International Expansion of US Manufacturing Firms   93

5.1 Introduction   93
5.2 The Significance of International Production and
Exporting as Means of Serving International
Markets   97
5.3 Changes in International Market Shares since
1974: the European and Japanese Corporate
Response to American Multinationals   102
5.4 The Sectoral Pattern of Success and Failure of
Countries by Comparison with their Firms   111
5.5 Summary and Conclusions   115

6 Technological Advantage as a Determinant of the
International Economic Activity of Firms   118

6.1 Introduction   118
6.2 Industrial Patterns in the Economic Advantages
     of Countries and their Firms   121
6.3 The Importance of Technological Advantage in
     the International Economic Activity of Firms   126
6.4 Conclusions and some Further Suggestions   133

Notes   136

7 Technological Competition and Intra-industry
   Production in the Industrialised World   137

7.1 Introduction   137
7.2 Some Evidence on Intra-industry Production and
     Trade in the Industrialised Countries   141
7.3 The Relationship between Intra-industry Trade
     and Production   148
7.4 Technological Competition between MNCs
     from West Germany and the USA   153
7.5 Conclusions   158

Notes   159

8 A Classical Model of the Impact of International
   Trade and Production on National Industrial Growth   160

8.1 Introduction   160
8.2 The Basic Model   162
8.3 The Impact of International Trade with
     Differential Rates of Innovation   167
8.4 The Impact of International Production with
     Differential Rates of Innovation   177
8.5 Conclusions and some Policy Implications   180

9 Towards an Evolutionary Theory of International
   Production   186

9.1 An Evolutionary Approach   186
9.2 A Survey of the Major Theories of International
     Production   188
9.3 Two Points at Issue between the Different
     Theories   206

Contents

9.4 The Development of an Evolutionary Analysis
of International Production   214

Notes   218

References   220
Index   233

Tables

2.1 A list of 30 leading European companies patenting in the US before 1914   22

2.2 The total number of US patents granted to residents of the major countries of origin   23

2.3 Indices of revealed technological advantage for the major industrialised countries in the periods (i) 1890–1912 and (ii) 1963–1983   26

2.4 The results of the regression of RTA in 1910–1912 on RTA in 1890–1892   32

2.5 The results of the regression of RTA in 1963–1983 on RTA in 1890–1912   36

2.6 The results of the regression of RTA in 1977–1983 on RTA in 1963–1969   38

2.7 The strength of the regression effect over the period 1963–1969 to 1977–1983   40

2.8 The results of annual and other sub-period regressions of the RTA index over the 1963–1983 period   41

3.1 The share of European industrial output accounted for by US majority-owned foreign affiliates, 1957–1977 (%)   70

4.1 The growth in exports of selected products: USA and Europe, 1955–1975   76

4.2 The growth in exports of European countries and the USA, 1955–1975   77

4.3 The sales of US manufacturing affiliates in Europe, and exports from the USA to Europe, 1957–1975 ($m)   80

4.4 The contribution of US affiliates in Europe to total European and US exports of selected manufactured products, 1957–1975 (%)   83

4.5   The share of majority-owned affiliates of US firms
      in the total exports of European countries,
      1957–1975 (%)   85
4.6   The growth of receipts and payments of royalties
      and fees from non-affiliate licensing 1965–1975
      (1965 = 100) of the USA and selected European
      countries   89
5.1   The geographical distribution of international
      production in manufacturing by source country, 1982   98
5.2   The geographical distribution of international
      production and indigenous firm exports in
      manufacturing combined, by source country, 1982   100
5.3   The industrial distribution of international
      production in manufacturing, 1982   102
5.4   The industrial distribution of international
      production and indigenous firm exports in
      manufacturing combined, 1982   103
5.5   National firms' shares of international production
      by manufacturing sector, 1974 (%)   104
5.6   National firms' shares of international production
      by manufacturing sector, 1982 (%)   105
5.7   National firms' shares of international production
      plus domestic exports of manufactured goods,
      1974 (%)   106
5.8   National firms' shares of international production
      plus domestic exports of manufactured goods,
      1982 (%)   107
5.9   The results of the cross-industry regression of
      national firms' share of international production
      in 1982 on their share in 1974   108
5.10 The results of the cross-industry regression of
      national firms' share of international production
      plus domestic exports in 1982 on their share in 1974   109
5.11 National shares of exports by manufacturing
      sector, 1974 (%)   112
5.12 National shares of exports by manufacturing
      sector, 1982 (%)   113
5.13 The results of the cross-industry regression of the
      national share of exports in 1982 on the national
      share in 1974   113

5.14 The results of the cross-industry regression of
    national firms' share of international production
    plus domestic exports on the national share of
    exports in 1974   114
5.15 The results of the cross-industry regression of
    national firms' share of international production
    plus domestic exports on the national share of
    exports in 1982   114
6.1 The revealed technological advantage index for a
    12-industry distribution 1972–1982   123
6.2 The index of revealed comparative advantage of
    countries across a 12-industry distribution, 1982   124
6.3 The index of revealed comparative advantage of
    each country's firms, across a 12 industry
    distribution, 1982   125
6.4 The results of the regression of the RCA index of
    countries in 1982 on their RTA in 1972–1982   126
6.5 The results of the regression of the RCA index of
    each country's firms' production for international
    markets in 1982 on their RTA in 1972–1982   128
6.6 The results of the regression of the RCA index of
    each country's firms' international production in
    1982 on their RTA in 1972–1982   131
6.7 The results of the regression of each country's
    ratio of international production to indigenous
    firm exports in 1982 on the world ratio and on
    the country's RTA index in 1972–1982   133
7.1 The index of intra-industry production, 1982   144
7.2 The index of intra-industry trade, 1982   147
7.3 The results of the regression of the intra-industry
    trade index for 1982 on the intra-industry
    production index for 1982   150
7.4 The index of innovation creation relative to usage
    based on US company-financed R&D, 1974, and
    patents, 1976–1977   152
7.5 The index of the relative attractiveness of the USA
    for the international production of West German
    firms and of West Germany for the international
    production of US firms   155

Preface

It has become very topical to be concerned with trends in the competitiveness of the major industrialised countries and their firms, related to underlying strengths and weaknesses in their performance in research and technological innovation. This book sets out to analyse these issues more precisely. The book originated in work that I did for my Ph.D. thesis, which was completed in 1986. In starting on the thesis I began with the intention of examining the impact on European industries and firms of the growth of US multinationals in Europe. I had hoped to use evidence on the effect of their investments on local market structure and industrial concentration, as a measure of the change of competitive conditions facing local companies. However, it became clear to me that increases in technological competition between firms can just as easily be associated with rising industrial concentration as with falling concentration ratios. This happens whenever the industry leaders with the greatest market shares are also the most innovative and fastest-growing firms.

To analyse the process of competitive interaction between expanding firms and its effects required a theory of the industrial pattern of innovation of the firms of each country. This I found in the theory of technological accumulation, first clearly articulated by Keith Pavitt of the Science Policy Research Unit at the University of Sussex. Related to this, I have also found useful recent work on the statistical theory of cumulative processes of path-dependent development, particularly by Brian Arthur and his colleagues, whose contribution was drawn to my attention by Giovanni Dosi. I have attempted to integrate certain aspects of the theory of technological change as a cumulative process with an analysis of the development of industries and their internationalisation. The focus of the study is the evolution of technological competition between the manufacturing firms of the major industrialised countries, and consequent changes in the structure of international industries. The new emphasis on dynamic industrial competitiveness led me to work with data on the patenting activity and the technological advantages of these firms, in place of data on industrial concentration ratios.

It also led me to reformulate my ideas on the locational effects of the investments of multinational corporations by looking at the implications for research and innovation in host countries. In the book I propose models which look at the impact of international trade and production on the growth of local industries and firms. I have been helped in this by my interest in classical economic thought, which was fostered by my undergraduate studies in Oxford (most notably the excellent seminar organised by Andrew Glyn and Walter Eltis), and by my early postgraduate work at Birkbeck College, London (where even the core courses took an historical approach). The reader will find evidence of economic ideas that first emerged in the writings of Smith, Ricardo and Marx at various places in the book, particularly in the more theoretical chapters. Smith and Ricardo devised the first dynamic theories of the impact of international trade on national economic growth, while Marx provided an early theory of technological accumulation.

In developing an analysis of the growth of the multinationals and international production, I have been very fortunate to have worked on the book and on other related projects at the University of Reading. This has allowed me the benefit of regular discussions with John Dunning (my Ph.D. supervisor) and Mark Casson, who are both internationally recognised authorities in the economics of international business. They have each made incisive contributions to the literature, and have influenced the nature of my research programme. The book has especially gained from my collaborative work with John Dunning on statistics of international investment and production (published as *The IRM Directory of Statistics of International Investment and Production*, London: Macmillan,

1987). This has helped me to create estimates of international production for the firms of the major industrialised countries, which are an indispensable part of the statistical sections of the book.

In addition, there have been many other gains from working in a Department which has a deserved reputation as a centre of excellence for its studies on the multinational firm. It has brought me into contact with a wide variety of complementary research undertaken both within Reading and elsewhere. In this respect, a series of lively seminars on international investment topics in Reading, and the opportunity to attend numerous conferences and meetings at home and abroad have been especially helpful.

In working on the economics of technological innovation, I have benefited from work done by Keith Pavitt and his colleagues at the Science Policy Research Unit at Sussex, and by Nathan Rosenberg at Stanford University. Their insights into technological change, and the gathering of new and readily usable data sources at SPRU, have provided enormous potential for research in this area. The scope now exists for integrating treatments of innovation into dynamic economic models, and for using data on technological advantage in rigorous statistical analysis. In this area I have also found the work of Schumpeterian economists, most notably Chris Freeman (former Director of SPRU), helpful, and I gained a great deal from discussions at the Conference on Innovation Diffusion organised by Giovanni Dosi and Fabio Arcangeli in Venice in March 1986.

I gratefully acknowledge the financial support of the Commission of the European Communities, who awarded me a one-year research grant for the academic year 1984–5, and of the Economic and Social Research Council, who provided me with a two-year grant at the start of my study in Reading. The data on US patent counts from 1963 to 1983 used in the book were prepared with the support of the Science Indicators Unit, US National Science Foundation, by the Office of Technology Assessment and Forecast, US Patent and Trademark Office. However, any opinions expressed in connection with the use of these data are those of the author, and do not necessarily reflect the views of the National Science Foundation or the Patent and Trademark Office.

I have enjoyed helpful discussions on the work as a whole with John Dunning, Mark Casson and Keith Pavitt, and with Francesca

Sanna Randaccio of the University of Rome. Chris Freeman and Luc Soete made stimulating remarks on an earlier draft of Chapter 2, which was presented at a seminar at SPRU in 1986, and which have helped me to think through the issues raised elsewhere in the book more clearly. Richard Goodwin provided useful comments on chapter 8. Another former colleague at Reading, Charles Sutcliffe, provided me with valuable help in computer programming, which enabled me to construct an index of technological advantage for any number of consecutive years, and for any chosen degree of aggregation across industries. At various times I have received useful comments or ideas from Peter Hart, Tony Corley, Steve Nicholas, Alan Rugman, Allan Webster, Daniele Archibugi, Pier Carlo Padoan, Gunnar Hedlund, Bruce Kogut, and anonymous referees. I am grateful for the interest they have shown and for the assistance they have given me, subject to the usual qualifying remark that none of them is to be held responsible for the final contents of the book, or for any errors that remain. I must also thank Mrs Yvonne Penford, who typed the first draft of the book on to the word processor system which I have been using, and who had to learn how to use the necessary software package in order to do so. Finally, I am grateful to those who have assisted me through the final stages of publication at Basil Blackwell: Mark Allin (Economics editor), Tracy Traynor (desk editor) and Peter Whatley (freelance copy editor).

Before beginning it may be useful to make explicit the procedure followed in designing the structure of the book. In the early chapters propositions are formulated and an analytical framework established from which certain predictions are derived. These are then tested, empirical evidence is considered on issues in general in the chronological order in which they have become important, results are presented and conclusions drawn. The findings are then used as a basis for constructing a theoretical model, and as a means of criticising and extending prevailing theories of international production, and suggesting directions for future research. The structure of the book may have the advantage that the reader is not required to be an expert in the field before beginning, but can read the book through from the start more easily.

1

Introduction

1.1 The Analysis of International Technological Competition

The post-war period has witnessed a massive rise in international trade and production in the industrialised world.[1] Whether in the phase of rapid growth in the 1950s and 1960s, or amidst slower growth rates since the mid-1970s, this form of organising economic activity and its major agent, the multinational corporation (MNC), has continued to expand relative to domestic production for domestic markets. Manufacturing industries have been at the forefront of this process of internationalisation. A series of international oligopolies have been created, characterised by differing degrees of rivalry between the leading MNCs involved. Each MNC retains its global position through a combination of technological innovation, and investment in a variety of complementary assets, most notably marketing and distribution networks.

This book sets out to examine the post-war evolution of technological competition between the manufacturing MNCs of the leading industrialised countries. Competition was at first strongest between US and European MNCs, but it has now widened to include fast-growing Japanese MNCs. The story begins after the formation of the EEC in 1958, when US manufacturing firms mounted a sustained wave of investment in Europe, in an attempt to preserve the strong position in European markets for research-intensive goods that they had held since the war. In the 1960s this precipitated a widespread concern in Europe at the 'technological gap' with the USA, and led some European commentators (such as Servan-Schreiber, 1967) to fear that significant sections of European industry would simply be taken over by US firms. However, by the 1970s, similar concerns about foreign technological competition were felt by the Americans themselves, as European firms began a counter-invasion of the USA. Today, in both Europe and the USA, the focus of discussion is on the activities of technologically advanced Japanese companies, although the Japanese are also worried about their performance in critical technological areas (see Patel and Pavitt, 1986). In judging the likely longer-run impact of Japanese investment in Europe and the USA, it is important to have understood the experience of the European response to the 'American Challenge'.

To appreciate more precisely the impact of inward investment on the development of a recipient industry and its firms, the book considers evidence on the impact of the US expansion in Europe on the competitiveness of indigenous firms. The earliest theories of direct investment were divided between those that argued that it necessarily had some anti-competitive impact on host country firms (for example, Hymer, 1976), and those that believed that its predominant effect was as a helpful competitive stimulus (for example, Kindleberger, 1969, and Knickerbocker, 1976). Such theoretical approaches are much too restrictive. An appraisal of the evidence on the response of European firms to US inward investment, as discussed in Chapter 5 below, should convince the reader that the response was uneven, and that it varied between countries and industries. Consequently, either a rise or a fall in indigenous technological activity is a possible outcome, and it still remains to determine the circumstances under which each of these types of effect comes to prevail.

Where the literature has addressed this question previously it has typically done so from the perspective of industrial organisation theory; that is, it has looked at the impact of inward direct investment on host country market structure and industrial concentration.[2] The underlying rationale is that changes in market structure affect the conduct and performance of indigenous firms operating in the market. The difficulty here is that market structure may not be a suitable guide to the extent of competitive pressure felt by local companies, or their ability to respond. Industrial concentration may be rising at the same time that technological rivalry between the leading firms in an industry (which are growing together) is on the increase. In order to capture this kind of effect it is necessary to shift away from the approach of 'industrial organisation', and towards that of 'industrial dynamics'

(Carlsson, 1987); that is to move the emphasis of interest away from the structure of an industry at a given point in time, and towards its evolution as a process over time at the international, national and firm level. In this context, the issue becomes whether firms in a given country and industry, when faced by the direct entry of foreign competitors, have the capacity to generate technological and allied advantages of their own that enable them to respond effectively, and in so doing to establish international networks of their own.

Technological competition between firms in an international industry can be thought of as an evolutionary process, in which the most successful companies become MNCs. In a given industry, firms that emanate from a country in which a strong innovative tradition has been established are likely to respond favourably to increased international competition. Such firms will have acquired a great deal of practical experience in the industry, and they are geared up to problem-solving activity (through fundamental research, production engineering, organisational change and so on) in order to implement relevant technologies in a satisfactory way. This gives them the capacity to generate fresh technological advantages, in part through assimilating and adapting complementary foreign technologies. If new technological developments arise in some foreign country, whose firms then embark on a course of MNC expansion, home country firms are well placed to catch up, and to begin to construct networks of international trade and production of their own. They in turn may eventually want to establish foreign production facilities in other countries which are themselves homes of strong innovative activity. If so, fundamental research work is liable to be increasingly concentrated in innovative centres in what has become an international industry.

The early chapters of the book lay the foundations for what follows by setting out the case for thinking of technological innovation as an evolutionary process, and investigating the consequences for competition in international industries. Most manufacturing MNCs, many of the strongest of which originally grew up towards the end of the last century, began as fairly technologically specialised firms before they gradually broadened the scope of their innovation. The relatively narrow range of specialisation of firms, and an international trend towards industrial concentration and collusion, led to the spread of international cartels and agreements between firms. It was only in the post-war period that international competition rather than cooperation between MNCs became the general rule (see Cantwell, 1988a). After 1945 the closely allied generation and use of technology by firms became increasingly linked to their creation of international networks of production.

The industrial composition of the expanding production networks of MNCs is conditioned by the sectoral pattern of their innovative activity, which regulates their ability to compete in world markets. A statistical study of the evolution of industrial innovation by the firms of the major industrialised countries appears in Chapter 2, which tests the extent of continuity in the sectoral composition of technological change for each such group of firms. If it is true that technological change is an evolutionary process in which innovation builds upon previous achievements, then the technological strengths of the firms of each of the industrialised countries are likely to shift only gradually over time. Using a procedure first developed by Soete (1980), a cross-industry index of technological advantage for the firms of each country is drawn up, based on their record in patenting activity. The discussion includes an historical perspective, using data compiled directly by the author from original patent records. The technique of Galtonian regression is used to examine the stability of the index of technological advantage, which is also new to research in this area. It has been developed and applied previously in the analysis of the size distribution of firms, and in the study of income distribution (Hart and Prais, 1956; Hart, 1976).

To investigate the progress of technological competition between the MNCs of the major industrialised countries, the analysis proceeds from what is known about the course of their innovative activity, rather than the industrial structure of the markets in which they operate. The European response to the American Challenge is examined in the light of the specialised technological capability of the firms of each of the industrialised nations. This is done in Chapters 3 and 4, which demonstrate the usefulness of understanding the technological evolution of companies when assessing the consequences of an increase in international competition, brought about through the continuing internationalisation of business. It is suggested that in a sector that is internationalised, innovation and industrial competitiveness tend to rise in countries in which indigenous firms have a tradition of technological strength, but may be weakened in countries in which domestic research and development is relatively limited.

Since a study of the competitiveness of national groups of firms over time requires an assessment of their overall shares in world markets, it is necessary to look not only at the position of their countries in international trade, but also at the growth of their own networks of international production. This is done in Chapter 5, which sets out new data on the world market shares of national groups of companies, and how they have been changing over time. The data on international production are entirely original, and are derived from a wide variety of sources by the author. The novelty in the approach comes not only from the incorporation of international production as well as trade in the total international economic activity of firms, but also by the revision of trade data to distinguish the exports of domestically owned firms from the exports of foreign-owned affiliates (which are part of the international production of the firms of other countries).

Drawing on this same new evidence, Chapter 6 returns to a discussion of the significance of technological advantage to the process of European firms catching up their American rivals in the post-war period. However, since the most competitively able European companies went on to expand their own MNC networks, their achievements must ultimately be judged not just on the recapturing of market shares within Europe, but more importantly on their market shares within their international industry. To this end, the industrial distribution of the value of the exports *and* international production of the firms of each major industrialised country are the relevant measure of comparative success and failure of national groups of firms. Moreover, when considering the 1970s and 1980s rather than the 1950s and 1960s, it becomes essential to include an assessment of the role of Japanese firms in the development of technological competition. The statistical work in the chapter considers the relationship between the industrial distribution of the total international economic activity of the firms of each of the major industrialised countries (exports plus international production), and the pattern of technological advantages they have been able to generate.

Chapter 7 extends this discussion to consider the implications for the structure of international industries. If technologically advantaged firms each increase the production and research that they locate in the international centres of innovation that are homes to their major rivals, then this will generate a steady rise in what is termed intra-industry production. Such intra-industry production is established through cross-investments by MNCs in the same industry. Other explanations of intra-industry production are possible, for example an increase in protectionism by each host country government. The chapter therefore examines the significance of technological competition as a driving force for the creation of networks of intra-industry production in modern manufacturing.

The book concludes by assessing the consequences of its findings for the theory of the impact of the internationalisation of production on domestic economic growth and industrial competitiveness. Chapter 8 advances a model of locational agglomeration in innovative activity which it suggests is applicable in the international industries that have grown up recently. It is a model of cumulative causation, in which a virtuous circle of cumulative success in one location is the counterpart of a vicious circle of cumulative failure in another. The model builds upon and develops the arguments of previous chapters. Its appearance towards the end of the book may be thought of as representing an historical as well as a logical sequence, in that the expansion of US firms in Europe gave way to a more general internationalisation, which in its turn has helped to create stronger international competition between alternative locations. Likewise, the final chapter ends by considering the implications of the study for future work on MNCs and the evolution of international economic activity, through a criticism and extension of the existing literature on MNCs. It compares the approach of the book to technological innovation and the growth of MNCs with other recent approaches, and suggests that the focus of attention should be shifted away from the structure of markets for industrial goods or the market for technology, and towards the conditions for successful research and technology creation.

1.2 The International Accumulation of Technology and Capital

The post-war growth of international trade and production of the manufacturing firms of the industrialised countries can be thought of as an accumulation of technology and capital within firms, and in international networks of production and trade. MNC expansion can be linked to a process of cumulative technological change within the firm, in which innovation and the growth of international production are mutually supportive. Having discussed the nature of the relationship between technological accumulation and capital accumulation in general terms, the book proceeds to analyse in more detail the extent to which innovation has underpinned the spread of international trade and production in manufacturing, and to look at the implications of rising MNC activity for the industries of the major industrialised countries and their firms.

The theoretical roots of the analysis can be traced back to the approach of the classical school of political economy (most notably the contributions of Adam Smith, David Ricardo and Karl Marx). According to them, the mainspring of a capitalist economy is the process of capital accumulation. Now in the case of the expansion of manufacturing industry, which has been central to capitalist development since the time of the industrial revolution, capital accumulation has been bound up with technological accumulation. The introduction of machinery enabled scientific advances to be applied in production (Rosenberg, 1982), and unleashed historically unprecedented rates of growth of production through progressive technological development. This, in turn, has been a major factor in the critical role played by manufacturing industry in economic development as a whole. Manufacturing industry is a substantial net creator of innovations, which are taken up in the extractive and service sectors of the economy (Scherer, 1984).

The term 'technological accumulation', originally coined by Pavitt (1987), encapsulates the idea that the development of technology within a firm is a cumulative process. That is, the creation of new technology is to be understood as a gradual and painstaking process of continual adjustment and refinement, as new productive methods are tested and adapted in the light of experience. In any firm there is a continual interaction between the creation of technology and its use in production. For this reason, although a group of firms in a given industry are likely to have similar lines of technological development (similarities which may be increased through collaborative R&D projects, through drawing on the results of publicly funded research, and through imitation), the actual technological path of each is to some degree unique and differentiated. The acquisition of new skills, and the generation of new technological capacity, partially embodied in new plant and equipment, must be a goal of every firm in an oligopolistic industry, if it is to maintain and increase its profits. Even where new technology is acquired from outside the firm, it must be gradually adapted and integrated with its existing production methods.

The notion of technological accumulation is consistent with the ideas of Rosenberg (1976 and 1982), Usher (1929), and the earlier work of Marx on technological change through systematic adaptation. More recently, Atkinson and Stiglitz (1969), Nelson and Winter (1977) and Stiglitz (1987) have spoken of 'localised' technological change in the context of the previous technological evolution and learning experience of the firm. The advantage of using a classical terminology in speaking of technological and capital accumulation, rather than simply innovation and investment, is that it emphasises that they are both continuous processes, and not just a series of discrete actions. It also calls attention to the way in which they are interlinked with one another.

In other words, economic growth consists of two components: on the one hand, the accumulation of capital equipment, associated in the longer run with the employment of an increasing labour force; and, on the other hand, the development of technology which increases the productivity of this capital and labour. Technology is both embodied in new items of capital equipment, and disembodied in improvements in the way it is used. Hence, technology is here defined with reference to the production process as a whole, and encompasses productivity improvements that are due to both scientific and organisational factors. A similar notion of technology can be found in Caves (1982), that it involves 'knowledge about how to produce a cheaper or better product at given input prices or how to produce a given product at lower cost than competing firms'.

This entails a deliberately broad definition of technological innovation to cover everything within the production process itself that, over time, raises the productivity of inputs. The phrase 'over time' is used to signify that innovation does not include productivity growth that is due purely to changes in the scale of output or the size of plant, using a given technology at a given point in time. The gradual improvement or accumulation of technology brings dynamic rather than static economies of scale, associated with changes in the conditions of production. Technology on this definition encompasses organisational capacity and managerial skills, as well as research and development, but excludes marketing (or at least that aspect of marketing that has to do with advertising, as opposed to the ability to differentiate products). However, for the largest MNCs of the industrialised countries there is likely to be a close relationship between the capability of a firm in fundamental research, and the way in which it organises production itself. In the empirical study that follows technological achievement is therefore reasonably measured by the research-based component of innovation (which gives rise to patenting activity), rather than its organisational aspects. The accumulation of technology in any case involves the gradual building up of largely intangible assets, and is reflected in the skills of the work force and the design of capital equipment.

Technological and capital accumulation run alongside one another, and the relationship between the two is so close that one cannot occur without the other. The international accumulation of technology and capital is one aspect of this combined process, and one that in recent years has become predominant. Technology accumulates in part through new productive experience, new skills, know-how, organisational capacity and so forth, but this is in general linked to the installation of new items of capital equipment, without which the flow of intangible improvements would soon slow down. Moreover, as technological experience and skills are developed in a particular area of production, existing capital equipment is adapted and updated in accordance. Technological change is neither of a purely disembodied, nor a purely embodied kind.

Many of the existing theories of international production have supposed that typically a parent MNC begins with an individual act of technology creation which is then diffused abroad through the operations of its foreign affiliates. This is not the case in the technological accumulation approach, in which the use of technology in new environments feeds back into fresh adaptation and (depending upon the state of local scientific and technological capability) new innovation. When production is located in an area that is itself a centre for innovation in the industry concerned, the firm may gain access to research facilities which allow it to extend technology creation in what are for it previously untried directions. In recent years technological accumulation has frequently been organised in such international networks, or in other words integrated MNCs. At one time MNCs may have been simply the providers of technology and finance for scattered international production; today they have become global organisers of economic systems, including systems for allied technological development in different parts of the world.

The technological accumulation approach therefore addresses the question of why it is that technology is developed in international networks, rather than in a series of separately owned plants. Part of the answer is provided by internalisation theory, which focuses on why MNCs as opposed to purely national firms have come into existence (see, for example, Buckley and Casson, 1976; Williamson, 1975; Teece, 1977; Caves, 1982). That is, if the initiating firm is to appropriate a full return on its technological advantage, and if it is to coordinate the successful introduction of its new technology elsewhere, then it must exercise direct control over the network as a whole. However, this may be not so much a feature of the market for technology which is the focus of internalisation theory, as a feature of the very nature of technological development itself.

Suppose for a moment that the act of exchanging technology between firms does not present a problem, in that a reasonable price for such an exchange can always be readily agreed. Now consider an international industry in which constituent firms produce more or less identical products for the same international markets. However, each firm has its own quite specific process technology, derived from a distinct technological tradition (say, different chemical processes with a similar end result). In this situation, if technological accumulation is continuous in each firm, raising its productivity or lowering its costs along a given line of technological development, then no existing firm would abandon its existing pattern of innovation and buy in all its technology from a competitor. It would be far more costly, and perhaps even infeasible, for an existing firm to switch into a completely new line of technological development, by comparison with the costs of the potential seller of technology simply extending its own network. Some exchanges of technology between existing firms will take place, since alternative lines of technological accumulation in the same industry are often complementary to one another. However, where technology is bought in it must be adapted to and incorporated into an existing stream of innovation, and this adaptation becomes part and parcel of the on-going process within an established firm of generating its own technology.

In the case outlined, the retention of technology within each firm has little to do with any failure or malfunctioning of the market for technology, but everything to do with the close association between the generation and the utilisation of a distinctive type of technology within each firm. By extending its own network each firm extends the use of its own unique line of technological development, and by extending it into new environments it increases the complexity of this development. The expansion of international production thereby brings gains to the firm as a whole, as the experience gained from adapting its technology under new conditions feeds back new ideas for development to the rest of its system. For this reason, once they have achieved a sufficient level of technological strength in their own right, firms are particularly keen to produce in the areas from which their major international rivals have emanated, which offer them access to alternative sources of complementary innovation.

Although it is becoming increasingly popular to treat technological change as a cumulative process (see, for example, Dosi et al., 1988) it is worth emphasising that this contrasts with the approach of many other economists in their writings on technology. It has been common to treat technology as analogous to knowledge or information, having some of the characteristics of a public good (an example being Dasgupta, 1987, or in the MNC field, Buckley and Casson, 1976). In particular, it has been suggested that, once created, technology is easily used by competing firms (providing they gain access to the necessary information) or transferred between different locations; it is not inherently firm-specific or cumulative. In this view there need be no particular association between technology creation and use within the firm, but the creation and use of technology may be treated as two quite separate issues. They are linked only through the market for technology (whether internal or external), and not as argued above through the conditions of production and technology adaptation and creation.

Another fairly common starting point is to make technological progress a function of entrepreneurship. This tends to lead to an emphasis on the random characteristics involved in the sectoral pattern of innovation, associated with major industrial change rather than cumulativeness and incremental change (going back to Schumpeter's notion of creative destruction). The proposition that technological change is a cumulative process is tested against the role played by random fluctuations in Chapter 2.

The approach taken here may also be contrasted with the view that MNCs steadily monopolise and reduce competition in their industries (advanced for example, by Cowling and Sugden, 1987). It is argued instead that the general internationalisation of the process of technological and capital accumulation in manufacturing industry in the post-war period has given rise to an increased international technological competition. During and since the rapid economic growth that characterised the industrialised world between the mid-1950s and the mid-1970s, the principal form of competition between MNCs has been the further accumulation of capital and technology in the international networks of each. As firms have become larger and more industrially diversified, supported by wider ranges of technology, they increasingly came into competition with one another at an international level.

On the face of it, it might seem that such technological competition may have played a less prominent role in the period of slower economic growth since the mid-1970s. The rationalisation of networks and collusive agreements between them have become more widespread. For at least some of the more established MNCs, the emphasis has shifted away from the extension of international networks, and towards an improved international division of labour within each network, though this may include the sub-contracting of some operations to smaller firms.

Sylos Labini (1984 and 1989) discusses the relationship between technological change and the form of oligopolistic competition further. He suggests, following Freeman and Perez (1989), that the shift in technological paradigm[3] away from energy-intensive lines of production (such as steel, motor vehicles and chemicals), and towards production related to electronic technology, has been associated with a reduction in static scale economies and an increasing role for smaller satellite firms with larger MNCs directly controlling more narrowly specialised plants. In fact, it is sufficient simply to recognise that there has been a reduction in the overall rate of innovation (as measured by total world patenting activity), and hence in technological accumulation, to explain the rationalisation of networks. The older lines of technological development have encountered increasing difficulties, while the newer lines have yet to acquire widespread industrial application; as a result the number of patents granted per unit of R&D spending (the 'productivity' of research) has fallen.

However, the technological accumulation approach suggests two major reasons why the growth in international production has been associated with sustained technological competition between MNCs in manufacturing industries. Firstly, internationalisation has supported technological diversification since the form of technological development varies between locations as well as between firms. By locating production in an alternative centre of innovation in its industry the MNC gains access to a new but complementary avenue of technological development, which it integrates with its existing lines. By increasing the overlap between the technological profile of firms competition between MNCs is raised in each international industry, although so also are cooperative agreements as the number of technological spillovers between firms increases as well. Spillovers occur where technologies are created by a firm which lie outside its own major lines of development, but which may be of greater use within the main traditions of another firm.

Secondly, and partly because of the first factor, today there are a growing number of connections between technologies which were formerly quite separate. This greater technological interrelatedness has brought more firms, and especially MNCs, into competition with one another. These two elements have been associated with the growth of what are sometimes called 'technological systems' in MNCs (Dunning and Cantwell, 1989). Where MNCs in a competitive international industry are all attracted to certain centres of innovation to maintain their overall strength, then research and research-related production may tend to agglomerate in these locations (Cantwell, 1987).

Evidence is considered in Chapters 6 and 7 which suggests that amongst the strongest industrial MNCs technological competition has remained important, even if certain types of technological cooperation have been on the increase. Even with a slower rate of expansion of MNC networks there is still scope for an increase in technological competition, due to an increasing degree of interaction between MNCs in international industries. What is more, for the firms of countries such as West Germany and Japan, MNC networks have been expanding even faster than in the past, in large part as a means of catching up lost ground on the longer-established US and British MNC rivals. Moreover, it remains possible that a new wave of innovation (based on electronics and biotechnology) may displace rationalisation and collusive agreements, and give rise to a fresh stream of more rapid technological and capital accumulation.

Note finally that the emphasis on technological accumulation and competition, and the rejection of the arguments of the critics of MNCs of the market power or monopoly capital school (examined further in Chapter 9) does not imply a necessarily benevolent view of MNCs. The book does not enter into the debate about whether MNCs are a 'good' or a 'bad' thing, but simply addresses the issue of why they have in fact grown in the way that they have. It aims to describe and explain the course of development taken by the firms of the largest industrialised countries in the post-war period, and how this development has been associated with certain patterns of wider industrial growth.

Periods of high rates of technological accumulation in the industrialised countries have been associated with faster capital accumulation and growth of output, and consequently higher real incomes. The major modern form of such technological and capital accumulation is that which takes place within international networks of production and trade. As a result, technological competition in international markets between the leading MNCs of the major developed countries has become a widespread rule in industries today, even at times when technological accumulation and output growth are relatively low. However, the location of activity in an international industry and the competitive standing of firms of different nationalities will depend upon the relative strength of innovation amongst the firms of each country. This issue is explored in an historical context in Chapter 2.

NOTES

1 International production is defined as production under common ownership, and hence some degree of common control across countries.

2 See, for example, Dunning (1974), Caves (1974), Rosenbluth (1970), Knickerbocker (1976), Globerman (1979) or Fishwick (1982); and, in the case of less developed countries, Lall (1978), or Newfarmer (ed., 1985).

3 A technological paradigm is defined as a widespread cluster of innovations which represent a response to a related set of technological problems, based on a common set of scientific principles and on similar organisational methods (see, for example, Dosi, 1984, and Freeman et al., 1982). Long wave theorists who adhere to the Schumpeterian tradition argue that the characteristics of the technological paradigm associated with each major upswing in innovative activity (such as that of the earlier part of the post-war period) are distinctive. This helps them to account for changes in technological leadership between upswings, as the organisational methods associated with different technological paradigms are likely to require the support of a different set of social institutions. This is also used to help them explain the delay in taking up a new technology paradigm until appropriate social changes can be brought about, and hence to explain a period of stagnation before the beginnings of a new innovative upswing are created. Technological accumulation has a strong tendency to remain within the prevailing technological paradigm, which establishes the boundaries within which problem-solving activity normally takes place.

2

Historical Trends in International Patterns of Technological Innovation

2.1 Introduction

This chapter tests the extent of continuity in the industrial pattern of innovation, as suggested by the theory of technological accumulation. This theory has been most clearly articulated by Pavitt (1987), although it is implicit in other recent literature (most notably Rosenberg, 1982, and Dosi, 1984). Pavitt argued that technology is firm-specific, cumulative and differentiated, and consequently that the industrial composition of innovative activity in a given location or amongst a given national group of firms reflects past technological accumulation. This suggests that international patterns of technological advantage, having been established, will remain relatively stable over time. The sectors in which each group of firms is technologically strongest change only gradually.

For these purposes, the theory of technological accumulation may be divided into three connected but separate propositions. The first is that technological change is cumulative, such that the sectoral composition of innovation amongst the firms of an industrialised country is by and large stable over periods of 10 or 20 years. It is this proposition which is statistically tested here, against the alternative proposition that technological change follows a random course, in which the relative strength of innovative activity is likely to regularly switch between industries. In association with these statistical tests, changes over time in the degree of technological specialisation of each national group of firms are examined using a related statistical procedure. The degree of technological specialisation is a measure of whether the innovation of the firms in question is highly concentrated in a few industries, or more broadly spread across a wider range.

It should be noted that the first proposition refers to the cumulativeness of technological development in terms of the industrial composition of innovation, not in terms of its overall rate or rapidity for each national group of firms. Why it is, for example, that West German or Japanese firms have on average enjoyed a faster rate of innovation than other firms, or why their average position has steadily improved, is not the issue here. The issue is the nature of the comparative advantage held by each national group of firms in technology creation, and the stability of that pattern of comparative advantage over time. This may, of course, have some bearing on which national group innovates most rapidly if technological activity rises fastest in electronics and Japanese firms are comparatively advantaged in that sector (see also note 3 of Chapter 1).

The idea that technological change is cumulative has also been attracting attention in other branches of economics and economic history. The first proposition also borrows from recent work on the mathematical and statistical properties of path-dependent processes (Arthur, 1984, 1988; Arthur et al., 1987). Later in the book it will be argued that technological accumulation helps to underpin cumulative patterns of growth in the production as well as the research activity of multinationals.

The second proposition of the theory is that technological change develops incrementally, so that firms tend gradually to move between related sectors. Although the underlying technology and skills continue to build upon the past, the industrial applications may gradually change, a particularly extreme case of which may come about with the formation of new industries. This proposition calls for extensive historical research, and so it is not explored in any depth here, but it is simply noted that it is in line with the findings of Rosenberg (1976 and 1982). However, the proposition implies that the industrial composition of innovation amongst each national group of firms may shift over longer historical periods. Statistical tests are conducted to show for which groups of firms this happens and the extent to which it happens in each case. In addition, descriptive evidence is presented on how the sectoral pattern of innovative activity of particular national groups of firms has actually evolved both in recent years and historically.

The third proposition is that technological change is differentiated between firms and locations. That is, the path of technological development followed by a particular firm or in a particular location is distinctive and characterised by elements that are specific to that firm or location. This proposition itself is not examined here, as to do so thoroughly would require intra-industry and microeconomic evidence which lies beyond the scope of the book. However, the proposition is called upon to help explain the motivation of firms in establishing international production, especially where such production is located in some international centre of innovation, supported by investment in local research facilities. This idea has been outlined in Chapter 1 (Section 1.2), and it is used when examining cross-investments by MNCs in Chapter 7, and the impact of internationalisation on locational patterns of growth in Chapter 8. None the less, it is not essential to the early empirical chapters of the book.

Returning, then, to the first proposition the theory of technological accumulation emphasises the cumulative characteristics of innovation. It is argued that the day-to-day adaptation of technology, through an interaction between its creation within a firm and its use in production, has a more pervasive influence than the major technological breakthroughs which give rise to entirely new production processes. Even radically new technologies, once they move beyond the purely scientific and experimental stage, often rely upon or are integrated with earlier technologies in the course of their development. For this reason, innovation tends to gather a certain logic of its own through the continual refinement and extension of established technologies. As specific technological experience is accumulated, the further development of production within the firm throws up new requirements, which its research and engineering departments must try and meet. Improvements tend to set the stage for their own future problems, which compel further modification and revision through the adaptation of production by innovative firms. Until there is a new stream of innovations based on a different set of fundamental discoveries, firms at the existing frontier of progress tend to establish dynamic advantages over others in the same industries. This helps to explain why, for example, West German firms in the chemical industry have maintained a strong tradition for a period of at least a hundred years.

In any given industry, firms based in certain leading countries where a tradition has been established tend to push forward with a sequence of innovations conditioned by the prevailing 'technology paradigm'. These firms are geared up to problem-solving R&D and production engineering in areas of technology in which they have accumulated a wealth of practical experience. There is a strong interaction between those responsible for production, who are utilising the latest processes available as they become feasible, and the staff of the R&D departments. Such firms generate strong technological advantages, in part through the assimilation of the relevant features of complementary foreign technologies.

To test this view of the way in which innovative advantages are created and maintained by firms from particular countries, it is necessary to set the analysis in historical perspective. If it is true that technological progress tends to be continuous rather than discontinuous, then patterns of technological advantage should change only gradually over time. To discover over what period of time such change is likely to become significant, an index of technological advantage was calculated for selected years spanning a period stretching back into the last century. The index was drawn up for each of the major industrialised countries. For a cross-section of industries, the index measures the strength of innovative activity of the firms of each country, and it is constructed for the period 1963–83, and for selected years before 1914.

It is an index of 'revealed technological advantage', and is calculated in much the same way as the index of 'revealed comparative advantage', familiar from the literature on international trade (see Balassa, 1965). In this case the index measures comparative advantage in innovative activity rather than comparative advantage in trade. For the industrial sector of any country, its revealed technological advantage (RTA) is given by the country's share of US patents taken out by foreigners in that sector, divided by its total share of US patents due to non-US residents. Hence, when the RTA index assumes a value greater than one the country concerned is relatively advantaged in that sector, while a number less than one indicates that its firms are relatively disadvantaged. Such an index was first used by Soete (1980).

The revealed technological advantage in industry i for the firms of country j is thus defined as:

$$\text{RTA}_{ij} = (P_{ij}/\Sigma_j P_{ij}) \, / \, (\Sigma_i P_{ij}/\Sigma_i \Sigma_j P_{ij})$$

where P_{ij} is the number of US patents in industry i granted to residents of country j.

2.2   A Description of the Data

The suitability of patent data as a measure of technological advantage is now quite well documented (for a review of the literature see Pavitt, 1985). While it is true that some innovations are never patented, and that some patents either have little qualitative impact or are never used, this leads principally to systematic industry-specific and country-specific differences, as it seems that firms from the same sector in any country have a similar propensity to patent. Scherer (1983), using US patent data, found that most of the variation between firms in the propensity to patent was to be explained by the extent of their research effort (as measured by R&D expenditure). Once inter-industry differences are allowed for, patenting as a measure of innovative output is strongly correlated with a widely used measure of innovative input.

Allowing that for the firms of a given country certain inter-firm intra-sectoral differences in the propensity to patent do exist, it seems reasonable to conclude that their variance is systematically lower than inter-sectoral differences. It can then be hypothesised that on relatively large numbers, the propensity to patent of a given national group of firms cannot be expected to have any systematic bias as compared to a notional industry average. A similar formulation could be applied equally well in the case of intra-national group differences in the propensity to patent in a given industry. Here, however, this additional assumption is not required as the RTA index is examined separately for the firms of each country, as will become clear.

The RTA index is normalised for both inter-industry and inter-country differences in the propensity to patent. The use of foreign patent data is further supported by the findings of Soete and Wyatt

(1983), that there is a strong inter-country correlation between foreign patenting and R&D, and a strong inter-industry correlation between foreign and domestic patenting. Moreover, the USA as a host to foreign patenting represents an important market for firms from the countries under comparison, so that they regularly take out patents there. The USA grants a higher number of patents to non-residents than any other country. A highly significant correlation between national shares of R&D and patenting in the USA was observed by Pavitt (1982).

Data on foreign patenting in the USA, organised by industry and country of origin, has been compiled by the Office of Technology Assessment and Forecast (OTAF) for years from 1963 onwards. Before 1963 the *US Index of Patents* provides a list of all patents granted in the USA by alphabetical order of the patentees, showing their state or country of origin. A brief description of each patent follows, with a patent number, such that if this is inadequate the complete description can be consulted in a separate catalogue. Using this source, US patents granted to foreigners can be allocated to the relevant countries and industries, just as OTAF has done for recent years. For this chapter, patent counts were made for the years 1890–2 and 1910–12. Table 2.1 indicates the importance of patenting in the USA to a number of technologically advanced European firms that had interests in the USA before 1914.

It has to be admitted that the use of foreign patenting activity in the USA before 1914 in the assessment of technological advantage is likely to be much less reliable than it is when working only with recent data. For a start, the assumption that for the firms of a given country the propensity to patent innovations can be treated over large numbers of observations as varying only with industry-specific factors is much more questionable. Moreover, it cannot be assumed that before 1914 there was the same inter-industry correlation between foreign and domestic patenting that exists today. The importance of international activity, and of the USA as a market and a source of competition, may well have varied between firms in the same industries. Indeed, individual inventors, with no formal affiliation to any firm, played a more prominent role at that time.

For these reasons, historical comparisons must be treated with caution. However, it would be unfortunate if a valuable source of

Table 2.1   A list of 30 leading European companies patenting in the USA before 1914

Chemicals
Badische Anilin & Soda
   Fabrik (Germany)**
Bauer & Co. (Germany)
Bayer (Germany)**
Burroughs & Wellcome (UK)
Cassella & Co.
   (Germany)*
Deutsche Gold &
   Silber-Sheide-Ansalt
   (Germany)*
Geigy & Co. (Switzerland)*
Heyden (Germany)*
Kalle & Co. (Germany)*
Lever Brothers (UK)
Merck (Germany)*
Nobel (France)
Society of the Chemical
   Industry (Switzerland)*
Solvay (Belgium)*
United Alkali Co. (UK)*

Electrical equipment
Bosch (Germany)
Howard & Bullough (UK)*
Marconi (UK)*
Orentein (Germany)
Siemens & Halske
   (Germany)**
Smidth & Co. (Denmark)

Motor vehicles
Daimler Motor Co. (UK)*
Fiat (Italy)
Rolls (UK)

Rubber tyres
Dunlop (Ireland/UK)*
Michelin (France)*

Textiles
Courtaulds & Co. (UK)
Linen Thread Co. (UK)

Metals
Deutsche Metallpatronen
   Fabrik (Germany)
Metallurgische Gesellschaft
   (Germany)

---

*   Denotes granted over 10 patents.
**   Denotes granted over 100 patents.
Source: *US index of Patents*, various issues.

evidence on historical patterns of technological advantage were to be entirely ignored for the purposes of statistical analysis. The RTA index can be conveniently compared over long periods of time, providing it is recognised that it is likely to change due to an improved representation of innovative activity, as well as due to actual changes in the sectoral pattern of technological advantage.

Table 2.2 sets out the numbers of US patents granted to residents of the major countries of origin in 1890–2, 1910–12 and 1963–83.

Table 2.2   The total number of US patents granted to residents of the major countries of origin

Country of origin	1890–1892	%	1910–1912	%	1963–1983	%
USA	66,766	91.6	95,022	88.6	929,133	69.1
Non-US total	6,084	8.4	12,285	11.4	416,113	30.9
UK	2,145	2.9	2,970	2.8	55,028	4.1
West Germany	1,378	1.9	3,961	3.7	101,864	7.6
Canada	975	1.3	1,673	1.6	22,160	1.6
France	548	0.8	1,031	1.0	38,956	2.9
Austria(-Hungary)	198	0.3	439	0.4	4,560	0.3
Australia	147	0.2	284	0.3	4,072	0.3
Switzerland	139	0.2	310	0.3	23,733	1.8
Sweden	101	0.1	318	0.3	14,621	1.1
Belgium and Luxembourg	54	0.1	149	0.1	5,125	0.4
Ireland	44	0.1	37	0.0	309	0.0
Italy	31	0.0	175	0.2	13,299	1.0
Denmark	22	0.0	94	0.1	2,760	0.2
Netherlands	19	0.0	56	0.1	12,317	0.9
Spain	17	0.0	35	0.0	1,316	0.1
Japan	6	0.0	34	0.0	94,046	7.0

Source: *US Index of Patents*, various issues; OTAF, unpublished data.

The ten foreign countries (counting Belgium and Luxembourg together) whose firms were granted the most patents before 1914 are today still all amongst the top thirteen. They are the UK, West Germany, Canada, France, Austria, Switzerland, Australia, Sweden, Italy and Belgium and Luxembourg. They have been joined by Japan and the Netherlands, as shown, and by the USSR which is excluded from the analysis as it raises a quite different set of institutional considerations. When considering the most recent period it is also feasible to include the USA (for which RTA is calculated as US firms' share of all patents granted, rather than all patents granted to foreigners) and three other EEC countries (Denmark, Ireland and Spain). Portugal and Greece are excluded as they have not attained very high levels of patenting in the USA even in recent years.

It should be made clear that for the recent period (1963–83), the cross-industry variation in the RTA index for any country predominantly reflects the structure of national groups of firms (especially the largest firms including MNCs), rather than that country's domestic industrial structure. This is because in the data on patent counts used for 1963–83, patents were in general registered by and thus attributed to parent companies, irrespective of the location of innovations on which they were based; and because large firms account for a substantial proportion of total patenting activity in the USA. For this reason the analysis refers to national groups of firms (owned by residents of the country in question), rather than simply national patterns of innovation.

In order to calculate the RTA index for years before 1914, each patent granted to a foreign resident in 1890–2 or 1910–12 has been classified to an industry using the description provided by the *US Index of Patents*. Time and budget constraints prevented the extension of this patent count (which already ran to over 18,000 patents) to the entire period 1890–1914, or to an additional period in the inter-war years, which would have been the ideal.

Historical comparisons are reported below for the top ten foreign countries, each of which accounted for over 200 US patents in the years 1890–2 and 1910–12 combined. The relevant values of the RTA index, calculated across 27 industries, are shown in Table 2.3. For other countries the distribution of the RTA index for these early years would be heavily influenced by the small numbers of patents in any category. This is illustrated by the rise in the number of zeros in the index as the number of patents falls. Taking the index for 1890–1912, Italy with 206 patents had three industries with an RTA of zero, Belgium and Luxembourg (203 patents) had four zeros, Denmark (116 patents) eleven zeros, and Japan (40 patents) no less then eighteen zeros. Countries outside the top ten must therefore be excluded from the historical aspect of the study. It may well even be the case that the historical analysis should be restricted to the top three countries (West Germany, the UK and Canada), each of which had over 2,000 patents in the early period, and which were the only countries with no zeros in the RTA index for these years.

The RTA index can be calculated for any number of consecutive years. Here, it has been calculated for 1890–2, 1910–12, 1890–1912 (from a combination of 1890–2 and 1910–12), 1963–83, 1963–9 and 1977–83. The stability of the RTA index, and changes in the extent of technological specialisation, are examined over different time periods. Section 2.3 describes the statistical methodology used to achieve this end. Sections 2.4, 2.5 and 2.6 below discuss the results, each referring to trends in technological advantage over a different time span, and Section 2.7 discusses some conclusions that might be drawn.

2.3   The Statistical Methodology

The theory of technological accumulation would suggest that for the firms of any given country, the sectoral distribution of the RTA index is likely to remain fairly stable over time. This means that if the RTA index is calculated for a national group of firms at two different points in time, then these two sectoral distributions of technological advantage should be positively correlated with one another. However, since the nature of innovative activity will change gradually over time, the degree of correlation is likely to fall, the further apart are the two groups of years under consideration. Over longer periods of time firms on a cumulative path of development may still move across technically related sectors.

The relevant statistical methodology is the Galtonian regression model, a statistical technique devised for the analysis of bivariate distributions. This approach was originally applied to economic problems in the context of work on the size distribution of firms by Hart and Prais (1956), and other useful applications have since been developed by Hart (1976) and Creedy (1985) in investigating changes in income distribution in the UK, and by Sutcliffe and Sinclair (1980) in the case of the seasonality of tourist arrivals in Spain. To adopt a similar procedure, the correlation between the sectoral distribution of the RTA index at time t and at the earlier time $t - 1$ is estimated through a simple cross-section regression of the form

$$RTA_{it} = \alpha + \beta RTA_{it-1} + \varepsilon_{it} \qquad (2.1)$$

This is estimated for a particular country, and the subscript refers to industry i at time t. The standard assumption of this analysis is that the regression is linear and that the residual ε_{it} is stochastic and independent of RTA_{it-1}. This is valid if the cross-industry index at each point in time approximately conforms to a normal distribution. The regression line will pass through the point of means and in Figure 2.1, for convenience of exposition, this is illustrated for the case where the mean of each distribution

Table 2.3 Indices of revealed technological advantage for the major industrialised countries in the periods (i) 1890–1912 and (ii) 1963–1983

Sector	UK (i)	UK (ii)	Germany (i)	Germany (ii)	Canada (i)	Canada (ii)	France (i)	France (ii)	Austria (i)	Austria (ii)	Australia (i)	Australia (ii)	Switzerland (i)	Switzerland (ii)
1. Food Products	0.63	0.96	1.26	0.61	0.34	1.39	1.34	0.75	1.37	0.74	0.85	1.80	1.90	1.36
2. Chemicals, n.e.s.	0.70	0.90	1.68	1.18	0.27	0.62	1.22	0.99	0.83	0.50	0.43	0.56	1.01	1.85
3. Synthetic Resins	1.03	0.73	0.47	1.29	1.68	0.32	0.57	0.77	0.00	0.48	0.00	0.55	7.87	0.78
4. Agricultural Chemicals	1.21	1.11	0.28	1.06	1.47	0.49	0.50	1.18	1.17	0.62	0.00	0.98	0.00	2.05
5. Cleaning Agents	1.47	1.54	0.78	1.02	0.99	0.30	0.19	1.22	1.32	0.40	0.69	0.64	1.29	1.31
6. Paints, etc.	0.17	0.93	2.82	1.24	0.16	0.64	0.28	0.65	0.09	0.92	0.00	0.99	1.78	0.93
7. Pharmaceuticals	0.72	1.02	1.41	0.88	0.63	0.57	0.91	1.20	1.55	0.70	0.90	0.60	1.41	1.55
8. Ferrous Metals	1.11	0.81	0.90	0.82	1.03	1.01	0.78	0.93	1.14	2.58	1.99	1.54	0.59	0.61
9. Non-ferrous Metals	1.05	1.01	1.03	0.70	0.87	1.50	1.23	0.93	0.76	2.00	0.40	1.52	0.74	0.73
10. Fabricated Metal Products	0.87	1.10	1.01	0.93	1.45	1.53	0.78	1.10	0.97	1.26	1.12	1.66	0.58	0.85
11. Mechanical Engineering, n.e.s.	1.04	0.98	0.89	1.12	0.91	0.96	1.05	0.93	1.09	1.25	0.98	0.99	1.12	1.01
12. Agricultural Machinery	0.43	0.95	0.52	0.86	2.35	2.19	0.34	1.01	0.81	1.20	3.14	3.62	1.18	1.37
13. Construction Equipment	0.90	1.15	0.89	0.97	1.18	1.70	0.20	1.06	1.64	2.41	1.47	1.51	0.34	0.85
14. Industrial Engines	1.76	1.27	0.46	1.06	0.53	0.70	0.99	1.00	0.67	0.76	0.82	0.93	0.99	0.59
15. Electrical Equipment, n.e.s.	1.25	1.04	0.70	0.86	0.95	0.74	1.29	1.11	1.49	0.63	0.64	0.53	0.97	0.68
16. Transmission Equipment	1.02	1.10	0.97	0.90	0.69	1.18	1.29	1.17	0.00	0.47	1.28	0.51	1.60	0.91
17. Lighting and Wiring	0.90	1.17	1.38	0.94	0.40	1.78	1.40	1.13	1.81	0.35	0.53	0.97	0.50	0.64
18. Radio and TV Receivers	1.21	0.80	1.72	0.62	0.26	0.60	0.00	0.63	0.00	0.71	0.00	0.28	0.00	0.34
19. Motor Vehicles	1.09	1.02	0.70	1.20	1.56	1.04	1.35	1.15	0.56	0.80	1.15	0.93	0.89	0.65
20. Shipbuilding	1.53	1.30	0.74	0.64	0.72	2.19	0.99	1.29	0.00	0.12	1.11	1.90	0.00	0.47
21. Transportation Equipment, n.e.s.	1.05	1.44	0.62	1.07	1.53	1.00	1.16	1.35	1.12	0.93	1.24	0.80	0.21	0.76
22. Textiles and Clothing	1.34	1.17	0.85	1.18	0.90	0.48	0.94	0.89	0.86	0.44	1.83	0.80	1.26	1.64
23. Rubber Products	1.75	1.03	0.62	1.06	0.60	0.93	1.21	0.94	0.44	0.92	1.19	0.97	0.32	0.64
24. Non-metallic Mineral Products	0.97	1.27	0.99	0.85	1.19	0.84	0.71	1.11	1.89	1.15	1.10	1.13	0.83	0.52
25. Coal and Petroleum Products	1.47	1.48	1.15	0.66	0.17	2.77	0.89	1.32	1.39	0.58	0.00	0.75	0.00	0.25
26. Professional Instruments	0.96	0.84	0.96	0.94	1.02	0.71	0.94	0.84	0.97	0.87	1.14	1.09	2.43	0.94
27. Other Manufacturing	0.98	0.87	0.90	0.80	1.68	2.26	0.82	0.96	0.82	1.91	0.98	1.74	0.70	0.84

n.e.s. = not elsewhere specified.

Table 2.3 (Continued)

Sector	Sweden (i)	Sweden (ii)	Italy (i)	Italy (ii)	Belgium and Luxembourg (i)	Belgium and Luxembourg (ii)	USA (ii)	Japan (ii)	Netherlands (ii)	Denmark (ii)	Spain (ii)	Ireland (ii)
1. Food Products	0.92	0.95	0.44	0.71	1.80	0.87	1.08	1.06	2.32	2.86	1.72	3.53
2. Chemicals, n.e.s.	1.06	0.31	0.95	1.29	1.57	1.11	0.91	0.37	1.01	0.68	0.75	0.00
3. Synthetic Resins	2.02	0.03	0.00	1.98	0.00	1.72	0.92	1.29	1.10	0.07	0.12	1.71
4. Agricultural Chemicals	0.00	0.50	7.64	1.21	0.00	1.11	0.84	0.75	0.76	1.10	1.52	0.90
5. Cleaning Agents	2.65	0.75	1.43	0.72	1.45	2.50	1.01	0.79	1.38	0.58	0.30	0.66
6. Paints, etc.	0.14	0.30	0.00	0.94	0.92	1.02	1.01	1.28	0.91	0.31	0.32	1.43
7. Pharmaceuticals	0.58	0.62	0.63	1.50	0.00	1.24	0.79	1.02	0.69	1.39	1.37	0.00
8. Ferrous Metals	0.71	1.51	1.31	0.75	0.89	2.16	0.83	1.13	0.57	0.24	1.20	0.00
9. Non-ferrous Metals	1.02	1.09	1.92	0.62	0.23	1.51	0.88	1.18	0.49	0.32	0.55	0.00
10. Fabricated Metal Products	1.21	1.57	0.52	0.85	1.12	0.84	1.10	0.74	0.94	1.36	1.37	1.23
11. Mechanical Engineering, n.e.s.	1.19	1.29	1.15	1.20	0.87	0.92	0.96	0.31	0.83	1.17	1.29	1.19
12. Agricultural Machinery	3.48	1.52	1.31	0.73	0.33	1.49	1.10	0.35	1.79	1.31	1.01	3.62
13. Construction Equipment	1.41	1.74	1.53	1.06	0.78	0.72	1.04	0.46	1.22	1.21	0.87	1.70
14. Industrial Engines	2.26	1.13	1.22	0.60	0.99	0.18	0.87	1.14	0.46	1.07	1.08	0.84
15. Electrical Equipment, n.e.s.	1.05	0.34	1.44	0.76	1.35	0.56	1.01	1.31	1.76	0.81	0.51	0.69
16. Transmission Equipment	1.23	1.22	3.56	0.94	0.90	0.78	1.03	1.05	1.18	0.92	0.64	0.63
17. Lighting and Wiring	1.02	1.03	1.84	1.29	0.37	0.83	1.09	0.72	1.95	0.68	1.75	1.45
18. Radio and TV Receivers	0.00	0.32	3.39	0.64	0.00	0.58	0.92	2.18	2.23	0.72	0.36	0.00
19. Motor Vehicles	0.26	0.99	1.40	0.80	1.43	0.42	0.94	1.06	0.31	0.35	1.02	1.41
20. Shipbuilding	0.54	1.32	3.48	0.77	0.59	0.21	1.03	2.66	1.66	0.86	2.60	0.88
21. Transportation Equipment, n.e.s.	0.84	1.25	1.97	0.87	0.77	0.59	0.98	0.78	0.22	0.21	1.32	0.69
22. Textiles and Clothing	0.10	0.40	0.45	0.95	0.34	1.07	0.92	0.99	0.64	0.91	0.85	0.00
23. Rubber Products	0.33	0.71	0.00	1.17	1.08	1.27	1.01	1.15	0.92	0.96	0.67	1.34
24. Non-metallic Mineral Products	0.57	1.14	0.46	0.74	2.02	1.99	1.00	1.03	1.02	1.64	0.84	1.24
25. Coal and Petroleum Products	0.00	0.33	1.13	0.63	4.60	1.05	1.26	0.65	2.72	0.55	0.28	0.00
26. Professional Instruments	1.39	0.32	0.75	0.62	0.26	1.53	0.99	1.41	0.73	1.08	0.55	0.91
27. Other Manufacturing	0.62	1.54	0.38	1.10	0.82	0.84	1.13	0.81	0.63	1.53	2.41	2.29

n.e.s = not elsewhere specified.

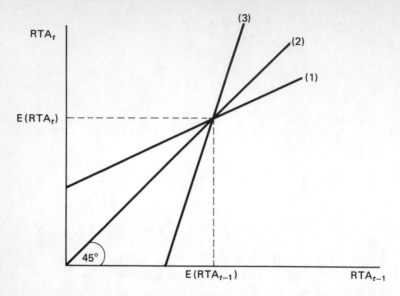

Figure 2.1   Galtonian regression with the RTA index

happens to be the same, the expected values of the distributions being given by $E(RTA_t) = E(RTA_{t-1})$. The analysis does not depend upon the values of the means being identical, but it can be seen that a sizeable difference between them may be indicative of a substantial measure of skewness in one of the indices, which consequently departs significantly from a normal distribution.

In Figure 2.1 the regression line (2) is drawn in such a way that the estimated coefficient $\hat{\beta}$ takes the value of one. This implies not only that the ranking of industries remains unchanged (advantaged industries remain advantaged, while disadvantaged industries remain disadvantaged), but also that they retain the same proportional position (advantaged industries do not become any more advantaged, and disadvantaged industries do not become any more disadvantaged). Where $\hat{\beta} < 1$, as in line (3), then there is a proportional shift in which already advantaged industries tend to become still more advantaged, while disadvantaged industries are increasingly disadvantaged.

In the case of regression line (1) where $\hat{\beta} < 1$ disadvantaged industries improve their position, and advantaged industries slip back. This is what has elsewhere been termed 'regression towards the mean' (Galton, 1889, cited in Hart, 1976). Where this is a true representation of $0 < \beta < 1$ then industries remain in the same ranking, but they come closer to one another. The magnitude of $(1-\beta)$ therefore measures the size of what is here called the 'regression effect', and this is the interpretation placed on the estimated coefficient $\hat{\beta}$.

In the case of $\beta < 0$ then the very ranking of industries would be reversed, contrary to the prediction of cumulativeness of the theory of technological accumulation. The expectation that $\beta > 0$, such that the RTA index is positively correlated across two points in time, can be readily tested for each country. The relevant test of $\hat{\beta}$ being significantly different from zero is the t-test. In a regression equation with only one independent variable the t-test is equivalent to the F-test, which refers to the significance of the correlation of the regression as a whole.

The test for whether $\hat{\beta}$ is significantly greater than zero is a test of the proposition of cumulativeness against the alternative that the sectoral composition of innovation is random. However, the second proposition to be set alongside that of cumulativeness in the industrial pattern of innovation is that of incremental change. If firms generally innovate in order to gradually adapt their existing technological strengths, they may still begin to shift the industrial nature of their activity. As the pattern of demand changes, and technology evolves, the sectoral distribution of innovation in a country may change, even though still drawing on a similar set of underlying technological skills. This effect is likely to be more pronounced the further apart the RTA distributions are in time.

The condition under which cumulativeness in the industrial distribution of innovation outweighs incremental change is that $\beta \geq 1$. Strictly speaking, if there were a path-dependent cumulative process with no change in the technological relatedness between sectors and therefore no shift in the underlying industrial structure of innovation (no incremental change), it would evolve towards a position where the proportion of innovations accounted for by each industry was stable and fixed (Arthur et al., 1987). This would correspond to $\beta = 1$, and to a regression effect $(1 - \beta)$ exactly equal to zero.

The test of whether cumulativeness outweighs incremental change in the period in question is hence the t-test of $\hat{\beta}$ not being significantly less than one (equivalent to a regression effect which is negative or not significantly different from zero). Where $\hat{\beta}$ is significantly greater than zero but significantly less than one then elements of cumulativeness and incremental change are combined. If cumulativeness dominates initially over relatively shorter periods ($\hat{\beta} \geq 1$), tests should reveal the length of time that it takes for incremental change to begin to play a significant role ($0 < \beta < 1$). What is then also required is that the regression analysis is supported by a more detailed inspection of the actual shifts in the RTA index, to investigate the actual evolution of sectoral strengths and weaknesses.

The other feature conveniently arising from the regression analysis of RTA distribution is a simple test of changes in the degree of technological specialisation. The degree of technological specialisation in a country can be measured by the variance of its RTA index, which shows the extent of the dispersion of the distribution around the mean. Pavitt (1987) used the standard deviation of the index, which is the square root of the variance, as a measure of such specialisation. The original work of Soete (1980) also analysed the variance of RTA indexes. The procedure for estimating changes in the variance of a distribution over time follows from Hart (1976). Taking equation (2.1) above, if the variance of the RTA index at time t is denoted by σ_t^2 then

$$\sigma_t^2 = \beta^2 \sigma_{t-1}^2 + \sigma_\varepsilon^2 \tag{2.2}$$

Now the square of the correlation coefficient (R^2) is given by

$$R^2 = 1 - (\sigma_\varepsilon^2/\sigma_t^2) = (\sigma_t^2 - \sigma_\varepsilon^2)(1/\sigma_t^2) \tag{2.3}$$

Combining equations (2.2) and (2.3) it follows that

$$\sigma_t^2 - \sigma_\varepsilon^2 = \beta^2 \sigma_{t-1}^2 = R^2 \sigma_t^2 \tag{2.4}$$

Equation (2.4) may be rewritten to show the relationship between the variance of the two distributions as follows

$$\sigma_t^2/\sigma_{t-1}^2 = \beta^2/R^2 \tag{2.5}$$

Hence the degree of technological specialisation rises where $\beta^2 > R^2$, and it falls where $\beta^2 < R^2$. A high variance indicates a high or narrow degree of specialisation, while a low variance indicates that the country has a broad range of technological advantage or a low degree of specialisation. Using the estimated regression values, the extent of specialisation rises where $|\hat{\beta}| > |\hat{R}|$, and it falls where $|\hat{\beta}| < |\hat{R}|$.

The estimated Pearson correlation coefficient, $\hat{R}$, is a measure of the mobility of industries up and down the RTA distribution. A high value of $\hat{R}$ indicates that the relative position of industries is little changed, while a low value indicates that some industries are moving closer together and others further apart, quite possibly to the extent that the ranking of industries changes. The magnitude of $(1-\hat{R})$ thus measures what is here described as the 'mobility effect'. It may well be that even where the regression effect suggests a fall in the degree of specialisation due to a proportional move in industries towards the average ($\beta < 1$), that this is outweighed by the mobility effect, due to changes in the proportional position between industries ($\beta > R$).

2.4 The Stability of Technological Advantage before 1914

The results of the regression of the RTA index in 1910–12 on the index in 1890–2 are reported in Table 2.4. Each distribution represented a cross-section of 31 industries for the ten major countries patenting in the USA before 1914.

Because of the problems created by a small number of observations, alluded to in Section 2.2 above, the critical assumptions required by the regression analysis are questionable for all but the top three countries. The major assumption is that the regression is linear, and this holds if the two distributions are both approximately normal. This was tested by calculating Pearson and Geary's measures of skewness and kurtosis for each distribution.[1] If the distribution were exactly normal then the test statistics for skewness (γ_1) and kurtosis (γ_2) would be zero. The values of γ_1 and γ_2 were calculated for 92 distributions (involving 15 countries over four time periods, 1890–2, 1910–12, 1890–1912, and 1963–83, and these 15 plus the USA for two periods, 1963–9 and 1977–83). To save space they are not listed separately, but those cases in which γ_1 or γ_2 are significantly different from zero, and hence the distribution significantly departs from the normal, are recorded in the text.

Outside the distributions of the UK, West Germany and Canada, the RTA index for the other 13 countries is significantly skewed in

Table 2.4   The results of the regression of RTA in 1910–1912 on RTA in 1890–1892

Country	$\hat{\alpha}$	$\hat{\beta}$	t_α	t_β	$\hat{R}$	β_1
UK	0.870	0.262	6.06**	2.13*	0.367	0.935
Germany	0.493	0.533	4.38**	5.24**	0.697	0.969
Canada	0.553	0.457	3.80**	3.41**	0.535	0.962
France	0.926	−0.073	9.23**	−1.75	−0.308	
Austria-Hungary	0.865	−0.004	5.36**	−0.04	−0.008	
Australia	1.067	0.009	1.07	0.01	0.009	0.790
Switzerland	1.066	0.252	2.91**	0.77	0.141	0.933
Sweden	0.726	0.047	4.75**	0.80	0.145	0.858
Italy	0.993	0.060	5.86**	2.92**	0.476	0.869
Belgium and Luxembourg	0.614	0.230	3.97**	2.81**	0.462	0.929

*   Denotes coefficient significantly different from zero at the 5% level.
**   Denotes coefficient significantly different from zero at the 1% level.
Number of observations = 31.

both 1890–2 and 1910–12, except for Austria-Hungary, Sweden and Belgium and Luxembourg in 1910–12. Also, apart from the cases of Austria-Hungary and Italy in 1910–12, the distributions are generally leptokurtic (narrow or tall) as indicated by significantly positive values of γ_2. By contrast, for the firms of the UK, West Germany and Canada the only significant measure of skewness occurs in the case of Canada in 1910–12, though it is still a great deal lower than for most other countries. The distributions for West Germany and for the UK and Canada in 1890–2 are leptokurtic, while those of the UK and Canada in 1910–12 are platykurtic (broad or wide, with $\gamma_2 < 0$), but not sufficiently to affect the regression analysis. The total number of patents required to construct an index that roughly conforms to a normal distribution obtained historically only in the cases of the UK, West Germany and Canada.

Accordingly, the results shown in Table 2.4 are only really trustworthy for the first three countries. In the case of France, which had the fourth largest number of foreign patents in the USA before 1914, the negative estimate of the $\hat{\beta}$ coefficient is indicative of the fact that the sectoral distribution for France in 1890–2 had a much higher mean value than in 1910–12. The mean value, and the skewness of the distribution, is increased by the RTA of the cleaning agents sector in 1890–2, which reached 11.54. This result arises due both to the relatively small number of French patents in the USA, and the relatively small number of all foreign patents in the cleaning agents industry in 1890–2.

If the basic assumptions of the regression are not met, then the t-statistics are not valid tests of the significance of the values of the estimated coefficients. However, for the three countries for which the results can be trusted with a reasonable degree of confidence, the findings appear to be broadly in line with expectations. The $\hat{\beta}$ coefficient assumes a positive value for all of these three, and indeed for eight out of all ten countries, demonstrating a positive correlation between the pattern of technological advantage in the early 1890s and just before the First World War. The t-test suggests that this correlation is significant in five of these eight cases. With reference to the three countries for which the data are most reliable, the correlation is significant for all three. Where the value of $\hat{\beta}$ was positive, the implicit value of β for a one-year regression was calculated as β_1 in the final column. This is calculated on the assumption that the structure of the relationship was unchanged throughout the period 1890–2 to 1910–12, and is the value of β that year-on-year would generate the observed $\hat{\beta}$ for a 20-year period. Although this procedure must not be taken too far, as the structure of the relationship was not unchanged (that is, the mobility effect or $1 - \hat{R}$ was greater than zero), it helps to illustrate that the value of the regression effect (measured by $1 - \hat{\beta}$) is very much weaker than would be suggested by the estimated value of $\hat{\beta}$ if considering periods of less than 20 years.

However, while for the three national groups for which there are a sufficient number of observations – British, West German and Canadian firms – randomness in innovation must be rejected in favour of cumulativeness, it is clear that the sectoral distribution of their technological development was also subject to a degree of change in the 20 years around the turn of the century. The regression effect was positive and significant ($\hat{\beta}$ was significantly less than one on a t-test) in all three cases. Although there is some evidence of incremental change during these years, it is difficult to draw conclusions at this level of analysis, given that many firms were only just beginning to establish themselves as US patenters at the time. This must itself account for a certain amount of change in the distribution of patenting activity.

For eight out of ten countries the value of $\hat{R}$ exceeded $\hat{\beta}$, signifying a fall in the degree of technological specialisation. That is, patenting activity in the USA tended to become broader in its sectoral scope. In part, this result may reflect an improvement in the representativeness of the data, and the number of US patents granted to foreigners rose substantially in the 20-year period in question (see Table 2.2). Once again, this also reflects the fact that many firms began patenting in the USA for the first time around the turn of the century. However, as the newer technologies pioneered at this time began to take root, there may well have been a broadening in the sectoral distribution of innovation in the leading industrialised countries.

The possibility that the sectoral distribution represented by the RTA index conforms more closely to a lognormal than a normal distribution was also investigated. The major reason for supposing that this might be the case is that while there is a lower bound to the distribution (disadvantaged industries are constrained to take RTA values between zero and one), there is no upper bound. Although there is some evidence which at first glance might support such a view, it in fact only strengthens the case against it. Distributions that are more closely lognormal than normal as judged by the tests of skewness and kurtosis are found only where the sample of patent counts is really insufficiently large, creating an artificially high degree of dispersion in the index. This is what tends to lead to a skewed distribution, rather than any inherent property of the RTA index itself.

Considering the five countries that were excluded from the historical analysis due to lack of data (Ireland, Denmark, the Netherlands, Spain and Japan), all have significantly skewed RTA distributions for both 1890–2 and 1910–12. However, in every case the value of the test statistic γ_1 is less than zero, indicating distributions in which the longer tail lies towards the lower values of RTA (to the left), which is a consequence of the large number of zeros. In other words, the theoretical justification for expecting a lognormal distribution is that there is no upper bound so it will be skewed to the right, but tests demonstrate the existence of lognormality only in instances where, due to the paucity of data, the distribution is skewed to the left. Therefore, although assuming a logarithmic rather than a linear functional form in such cases sometimes improved the value of the t-statistic on $\hat{\beta}$, and the estimated correlation coefficient, $\hat{R}$, these results are not discussed here.

It is worth noting those sectors in which the leading three countries (the UK, West Germany and Canada) were strongly advantaged in both 1890–2 and 1910–12 (see also Table 2.3). British firms appear to have maintained a strong technological advantage at this time in cleaning agents (mainly in soaps and detergents; the early French strength here mentioned above was in perfumes and cosmetics), industrial engines and turbines, ship-building, textiles, rubber products (tyres) and coal and petroleum products. German firms enjoyed a favoured position in chemicals in general, but especially in dyestuffs and paints, and in lighting and wiring equipment. Canada's strength lay in agricultural chemicals, railways and other transportation equipment, paper products, and other manufacturing (including furniture and wood products). In the case of these three countries at least, as far as can be told, the sectoral pattern of technological advantage remained fairly stable (elements of cumulativeness were present) in the 25 years before 1914, although the degree of specialisation fell.

2.5   The Stability of Technological Advantage over the past Hundred Years

Table 2.5 shows the results of the regression of the RTA index in 1963–83 on the index in 1890–1912 (the years 1890–2 and 1910–12 combined). These relied on the cross-section of 27 industries given in Table 2.3. Of the original 31 industries in the historical series, to ensure comparability with the recent data, drink and tobacco were subsumed under food products, leather products were subsumed under textiles, leather and clothing, and paper products included under other manufacturing.

In constructing the RTA index for the whole span 1890–1912, the West German distribution joined those that were significantly skewed away from normality, but those of Austria-Hungary and Australia improved (with γ_1 falling close to zero). In the case of West Germany the problem arises because of the industrial re-aggregation, which causes a number of industries whose RTA values were similar to be grouped together or lost information which was important historically is lost in moving to the more recent classification. The Austrian and Australian distributions are helped by the increased patent count pertaining in the combined

Table 2.5   The results of the regression of RTA in 1963–1983 on RTA in 1890–1912

Country	$\hat{\alpha}$	$\hat{\beta}$	t_α	t_β	$\hat{R}$	β_1
UK	0.803	0.256	6.97**	2.48**	0.445	0.981
Germany	0.957	−0.147	11.22**	−0.19	−0.037	
Canada	1.013	0.120	3.88**	0.50	0.099	0.971
France	0.912	0.127	10.41**	1.38	0.266	0.972
Austria	0.731	0.241	3.19**	1.13	0.219	0.980
Australia	0.537	0.632	3.27**	4.44**	0.664	0.994
Switzerland	0.901	0.026	8.02**	0.43	0.084	0.951
Sweden	0.804	0.163	5.50**	1.43	0.276	0.975
Italy	0.967	−0.016	11.03**	−0.40	−0.077	
Belgium and Luxembourg	1.066	0.125	6.66**	0.10	0.063	0.972

*   Denotes coefficient significantly different from zero at the 5% level.
**   Denotes coefficient significantly different from zero at the 1% level.
Number of observations = 27.

period, and by the industrial reclassification. However, the Canadian and Austrian distributions for 1963–83 were positively or rightwardly skewed, contrary to their closeness to normality in 1890–1912. Some difficulties with the interpretation of the linear regression analysis may again be anticipated for this reason.

A positive correlation between the two distributions was obtained for eight out of ten countries. However, on the whole the extent of this correlation was poor, and it was significant in only two cases. It is the firms of the UK and Australia that appear to have shifted least from the sectoral patterns of technological advantage that prevailed historically. In these cases as well, though, the regression effect is still significant ($\hat{\beta}$ is significantly less than unity). It is not terribly surprising that the industrial composition of innovation shifts substantially over long historical periods. The value of β_1 generally remains high, but the distributions are over 70 years apart, and it is rather unrealistic to suppose that the structure of the model has remained unchanged throughout this time.

The other noticeable feature of Table 2.5 is that the estimated value of $\hat{\alpha}$ is high for every country, and everywhere significantly different from zero. This might suggest a low value of β and a strong regression effect. However, it seems that what it actually represents is the weakness of correlation between the two RTA distributions, which is associated with a strong mobility effect. Leaving aside the two countries for which correlation was significant and the degree of specialisation fell, in four of the remaining eight countries the mobility effect outweighed the regression effect leading to an apparent rise in the degree of specialisation. For the firms of West Germany, Canada, Austria and Belgium a low magnitude of $\hat{\beta}$ was offset by a still lower value of $\hat{R}$.

Of course it might be that these results would be altered if patent counts were made over the full period 1890–1914, rather than for only six years. However, it seems unlikely that this would affect the conclusion that the sectoral distribution of technological advantage has tended to shift quite substantially since the turn of the century. It is possible to say a little on what might be the long-term effects of incremental changes in the composition of innovation. British firms seem to have moved from synthetic resins towards paints and pharmaceuticals, from ferrous metals towards metal products, and from general mechanical engineering and industrial engines towards agricultural and construction equipment. West German firms appear to have improved their position in mechanical engineering, particularly in industrial engines, and in motor vehicles and other transportation equipment. This might be related to traditional strengths in metal products and certain categories of electrical goods. The improvement of the West German group in synthetic resins, agricultural chemicals and cleaning agents may be similarly linked to their long-standing overall strength in general industrial chemicals. Shifts in the RTA index for other countries can be determined from an inspection of Table 2.3, though they are not as reliably indicated due to the influence of small numbers of patents in the earlier period.

2.6 The Stability of Technological Advantage since 1963

The estimates obtained form the regression of the RTA distribution in 1977–83 on the distribution in 1963–9 are set out in Table 2.6. They were derived from the same cross-section of 27 industries listed in Table 2.3.

Pavitt (1987) reported on a similar set of correlations for the regression of the RTA index for ten countries in 1975–80 on the equivalent RTA in 1963–8. He also used a 27-industry

Table 2.6   The results of the regression of RTA in 1977–1983 on RTA in 1963–1969

Country	$\hat{\alpha}$	$\hat{\beta}$	t_α	t_β	$\hat{R}$	β_1
USA	−0.513	1.513	−2.97**	8.72**	0.868	1.030
West Germany	0.341	0.642	2.66**	4.85**	0.696	0.969
Japan	0.586	0.339	4.22**	2.88**	0.500	0.926
UK	1.124	−0.216	4.15**	−0.08	−0.017	
France	0.758	0.306	3.71**	1.54	0.295	0.919
Switzerland	0.168	0.905	1.38	7.49**	0.832	0.993
Canada	−0.143	1.141	−0.77	7.64**	0.837	1.009
Sweden	0.130	0.979	0.79	6.11**	0.774	0.998
Italy	0.681	0.231	6.10**	2.36**	0.428	0.901
Netherlands	0.327	0.712	1.82	5.40**	0.734	0.976
Belgium and Luxembourg	0.806	0.401	2.82**	1.55	0.297	0.937
Austria	0.399	0.572	2.58*	4.55**	0.673	0.961
Australia	0.149	0.794	0.77	5.56**	0.745	0.987
Denmark	0.063	0.889	0.29	4.71**	0.686	0.992
Spain	0.384	0.855	1.43	3.30**	0.551	0.989
Ireland	0.636	0.543	2.01	2.85**	0.495	0.957

*   Denotes coefficient significantly different from zero at the 5% level.
** Denotes coefficient significantly different from zero at the 1% level.
Number of observation = 27.

disaggregation, though he adopted a rather different sectoral classification. The other important difference here is that the original patent data have now been reworked by OTAF, based on a 'fractional' allocation of those patents that had previously been allocated to more than one industrial group.

In terms of the measures of skewness and kurtosis, departures from normality were observed in the RTA distributions of Canada, Austria, Switzerland, Ireland, Italy and Denmark in both 1963–9 and 1977–83, for Sweden in 1963–9, and for the UK, Australia, the Netherlands and Spain in 1977–83. However, these variations from normality were in general much lower than those encountered in the historical series, and the problems created are correspondingly more limited. The only serious exception to this comes in the case of Ireland, whose firms had a low level of patenting activity even in the recent period, and which has a sizeable number of zeros in the RTA index for the latest years. As a result, the γ_1 measure of skewness for Ireland in 1963–9 reached the incredibly low magnitude of −67.6.

In his study, Pavitt found a positive and significant correlation between the two RTA distributions for nine out of the ten countries. He also found that the sectoral pattern of technological advantage of each country was distinctive, in the sense that there was little association between the distributions of any two countries. A similar state of affairs applies to the 16 countries listed in Table 2.6. A positive correlation holds for 15 of the 16 countries, and for 13 of these it is significant at the 5% level (it is significant for the other two at the 15% level). Only the UK demonstrates little relationship between the distributions characterising the two periods in question.

Unlike in the historical application of the analysis, the estimate of the α coefficient is negative in the case of two countries (the USA and Canada), though in the case of Canada it is not significantly different from zero. The corollary, given that the means of the two distributions are not too far apart, is that $\hat{\beta}$ exceeds unity, implying a negative regression effect, which is particularly strong for the USA. Indeed, for 11 of the 16 countries, despite generally high values of $\hat{R}$, the mobility effect (which is measured by $1-\hat{R}$) exceeds the regression effect (which is measured by $1-\hat{\beta}$). This means that there has been a tendency for the degree of technological specialisation to rise over the past 20 or 25 years. In other words, not only has the pattern of technological advantage remained fairly consistent, but advantaged industries have often acquired a stronger position than was the case for high-ranking sectors in the 1960s.

The strength of the regression effect over the period can be separately measured by testing whether $\hat{\beta}$ is significantly different from one (which amounts to a test on whether the regression effect, $1-\hat{\beta}$, is significantly different from zero). The results of this test are reported in Table 2.7. This shows that the regression effect was insignificant for half of the 16 countries. This includes the case of the USA, for which $\hat{\beta}$ was significantly different from one, but it was substantially above one, and hence the regression effect was significantly negative. US firms have become increasingly specialised in areas in which they were already advantaged. In the historical analysis the regression effect was everywhere significant, so this helps to demonstrate the way in which it gradually increases over longer periods. However, between the early 1960s and the early 1980s it fails to reach significant proportions for eight countries. Thus for half the national groups cumulativeness outweighs incremental change over this period.

Table 2.7   The strength of the regression effect over the period
1963–1969 to 1977–1983

Country	t_β
USA	2.96**
West Germany	−2.70*
Japan	−5.62**
UK	−4.73**
France	−3.49**
Switzerland	−0.79
Canada	1.07
Sweden	−0.13
Italy	−7.86**
Netherlands	−2.18*
Belgium and Luxembourg	−2.32*
Austria	−3.40
Australia	−1.44
Denmark	−0.59
Spain	−0.56
Ireland	−2.40*

*   Denotes coefficient significantly different from one at the 5% level.
**   Denotes coefficient significantly different from one at the 1% level.
Number of observations = 27.

As has been mentioned above, the calculation of β_1 from the estimated value of $\hat{\beta}$ over a longer period is unreliable as it assumes a zero mobility effect, which is known to be untrue. For the recent period the greater level of patenting activity makes it feasible for the firms of some countries to estimate the actual magnitude of successive values of $\hat{\beta}_1$ from year-on-year regressions. This was done for the ten countries whose residents were granted over 10,000 patents in the USA between 1963 and 1983, each averaging over 500 patents per year. The first column of Table 2.8 reports the average value of the estimated slope coefficients from a series of annual regressions for each of these ten countries. Alongside the mean of the annual $\hat{\beta}_1$ (denoted by $\mu_{\hat{\beta}_1}$), the standard deviation around the mean ($\sigma_{\hat{\beta}_1}$) is shown in the second column. The regression estimates of $\hat{\beta}_1$ are uniformly smaller than the calculated β_1 of Table 2.6, since the annual mobility effect does not

Table 2.8   The results of annual and other sub-period regressions
of the RTA index over the 1963–1983 period

Country	$\mu_{\hat{\beta}_1}$	$\sigma_{\hat{\beta}_1}$	$\hat{\beta}_7$(1970–1976/ 1963–1969)	$\hat{\beta}_7$(1977–1983/ 1970–1976)
USA	0.967	0.095	1.242	1.194
West Germany	0.799	0.121	0.746	0.854
Japan	0.884	0.107	0.456	0.857
UK	0.746	0.185	0.632	0.647
France	0.621	0.173	0.654	0.556
Switzerland	0.825	0.273	1.127	0.803
Canada	0.893	0.122	1.054	1.139
Sweden	0.789	0.222	0.872	1.149
Italy	0.539	0.168	0.500	0.454
Netherlands	0.744	0.165	0.730	1.008

always work in the same direction. This implies that for each national group of firms there is to some extent a random short-run fluctuation of industries up and down the ranking of innovative activity, in addition to any gradual but sustained shift in ranking that may be slowly taking place.

Of the groups of firms of all countries, Japan is the only one whose $\hat{\beta}_1$ over the period 1963–83 appears to have been on a significant upward trend over time. Regressing the time series of $\hat{\beta}_1$ on a simple time trend, the coefficient on the trend was positive and significantly different from zero at the 5% level for Japan, but insignificant for all other countries. Noting that Japan is one of the minority of countries whose firms have experienced a recent fall in their degree of technological specialisation ($\hat{\beta} < \hat{R}$ in Table 2.6), what this suggests is that the broadening of innovative activity by Japanese firms was especially strong early in the period.

Further support for this view emerges when dividing the 1963–83 period into three sub-periods (1963–9, 1970–6, and 1977–83), and running regressions for consecutive sub-periods. The estimated $\hat{\beta}_7$ coefficients from the (seven-year period) regression of the RTA index in 1970–6 on that in 1963–9, and of RTA in 1977–83 on that in 1970–6 are also given in Table 2.8. It can be seen that they are not tremendously different from the individual year-on-year $\hat{\beta}_1$, and that for most countries $\hat{\beta}$ does not change very much over the period as a whole. The only real exception as before is Japan, for which the value of $\hat{\beta}_7$ nearly doubles, confirming that $\hat{\beta}$

has been on something of an upward trend. It should also be noted that the value of $\hat{R}$ is very high for Japan in all the sub-period regressions, which confirms that Japanese firms have experienced a widening degree of technological specialisation rather than a major shift in the ranking of industries in terms of the strength of their innovative activity.

In this connection, what is also noteworthy from Table 2.6 is that the firms of those countries in which innovation, accompanied by industrial restructuring, has been proceeding most rapidly since the early 1960s, have still retained a fairly stable sectoral pattern of technological advantage. Japan, for instance, is not as might have been supposed a counter-example to the notion of technological accumulation, that is innovation through the further development of existing strengths. Japanese firms today retain an advantaged position that was already apparent by the 1960s in non-ferrous metals, electrical equipment in general, and electrical transmission equipment and radio and TV receivers in particular, rubber products, and professional and scientific instruments. They have witnessed an improvement in the innovative capacity in motor vehicles and industrial engines, and a decline in chemicals and shipbuilding; the traditional strength in shipbuilding seems to have led to innovation in metals and electrical equipment, and from there to motor vehicles.

The technological advantage of US firms has held up and been strengthened in fabricated metal products, lighting and wiring equipment, coal and petroleum products, and other manufacturing (which in the 27-industry case includes paper, printing and publishing). The position of ferrous metals, radio and TV receivers and motor vehicles amongst US firms has been weakened. West German companies have maintained a continual technological strength in chemicals, including synthetic resins, cleaning agents and paints and varnishes, and in motor vehicles, textiles, and rubber products. They have improved their advantage in certain areas of mechanical engineering, notably in construction and mining equipment, but have slipped back in radio and TV receivers and professional and scientific instruments.

The UK, which is not noted for its comparative innovative performance in the past 20 years, is the country whose firms have experienced the greatest 'mobility effect' in terms of the movement of industries around the RTA distribution. The UK industrial groups which have fallen furthest down the distribution are non-ferrous metals, mechanical engineering, radio and TV receivers, motor vehicles, and coal and petroleum products. The sectors in which UK firms have most improved their innovative standing are food products, agricultural chemicals, cleaning agents (though here there is some evidence that the RTA index was unusually low in 1963–9), pharmaceuticals, agricultural machinery, and non-metallic mineral products.

This suggests an interpretation which, if correct, would provide further support for the technological accumulation approach. Where a country is lagging in innovation and productivity growth, the industrial structure of the technological advantage of its firms may well be disrupted, as previously strong firms decline and perhaps even go out of business. Where, on the other hand, innovation and productivity growth is proceeding rapidly (as it has been in Japan and West Germany, at least until very recently), then this will in general tend to strengthen and reinforce an existing pattern of technological advantage. The strongest trends in innovation tend to be those that are grounded on a previously attained level of skills and experience in related technology. However, this argument cannot be pushed too far on the basis of the evidence considered here. A cross-national group regression of the measure of the mobility effect $(1 - \hat{R})$ on the proportional change in the overall share of patenting in the USA showed the expected negative correlation but it was not significant.

One other comment on the relationship between the industrial composition of innovation and overall technological performance can be made. This concerns the extent of representation of each national group of firms in the sectors of fastest-growing technological activity. For any country j denote the proportion of US patents held in industry i by p_{ij}, the share of total world patenting in the USA by w_j, and the mean value of the RTA index by M_j. That is (supposing there are altogether n industries):

$$p_{ij} = P_{ij}/\Sigma_i P_{ij}$$

$$w_j = \Sigma_i P_{ij}/\Sigma_i \Sigma_j P_{ij}$$

$$RTA_{ij} = p_{ij}/w_j$$

$$M_j = \Sigma_i p_{ij}/n w_j$$

The ratio of the mean value of RTA in an earlier period $t - 1$ to that in the next period t is then given by:

$$M_{jt-1}/M_{jt} = (w_{jt}/w_{jt-1})/(\Sigma_i p_{ijt}/\Sigma_i p_{ijt-1}) \tag{2.6}$$

Now since the regression equation (2.1) must pass through the point of means it is also known that:

$$M_{jt} = \hat{\alpha} + \hat{\beta}M_{jt-1} \tag{2.7}$$

This may be rewritten:

$$M_{jt-1}/M_{jt} = (M_{jt} - \hat{\alpha})/(\hat{\beta}M_{jt-1}) \tag{2.8}$$

From equations (2.6) and (2.8) it follows that where $(M_{jt} - \hat{\alpha}) >$ $\hat{\beta}M_{jt-1}$ that the world share of patents (w_j) rises faster than the average proportion held in industries at the chosen level of disaggregation $(\Sigma_i p_{ij}/n)$. It can be shown that this happens *either* because of a shift in the structure of industry proportions and thus in the cross-industry pattern of RTA (what has been termed above the mobility effect), *or* because the firms in question are particularly advantaged in industries with the fastest-rising patent numbers. Conversely, where $(M_{jt} - \hat{\alpha}) < \hat{\beta}M_{jt}$ there is either a mobility effect in the opposite direction,[2] or firms are relatively stronger in industries that have the slowest-growing technological activity.

Of the 16 national groups, there are seven for which the difference between $(M_{jt} - \hat{\alpha})$ and $\hat{\beta}M_{jt}$ for the most recent period was greater than 10%. In the case of two, the UK and Belgium and Luxembourg, this seems to be due primarily to a substantial mobility effect (measured by a high value of $1 - \hat{R}$ from Table 2.6). Of the other five, Japanese and Italian firms seem to have held their greatest advantage in industries with the fastest-growing innovation $(M_{jt} - \hat{\alpha} > \hat{\beta}M_{jt})$, while Swedish, Spanish and Irish firms were relatively advantaged in industries characterised by the weakest growth in technological activity $(M_{jt} - \hat{\alpha} < \hat{\beta}M_{jt})$. This confirms the suggestion that the improved performance of Japanese companies is partly attributable to their having held a comparative technological advantage in sectors which have been the leading sources of new increased innovation.

2.7 Conclusions and some Possible Extensions

The statistical evidence on international sectoral patterns of technological advantage offers support to the idea that innovation tends to unfold as a cumulative process, accompanied by gradual incremental change. The basis for this contention also seems to have become historically stronger. The modern firm relies even less on the external evolution of science and technology, and even more on the internal creation and refinement of new productive methods and new products. For this reason the sectoral distribution of innovative advantage of the firms of most technologically advanced industrialised countries tends to be correlated over time. However, due to a gradual shift in the nature of technology and the pattern of demand, this statistical link becomes tenuous over longer periods. The sectoral pattern of innovative activity gradually changes as new industries develop and new technological linkages are forged between sectors. Yet this is a slow process which in general only slightly disturbed the pattern of technological advantage held by the firms of the major industrialised countries in the 20 years between the early 1960s and the early 1980s.

It is clear that there may be applications of this theory of how firms tend to create innovative advantages in studies of technological competition in key industries. The evidence of this chapter suggests that countries and regions are likely to have their greatest scope for future innovation and growth in areas closely related to those in which their firms have been successful in the past. Thus, for example, Europe may well have the potential to exploit discoveries in biotechnology, given the traditional strength of her firms in the chemicals industry (see Patel and Pavitt, 1986). More precisely, joint ventures on a European-wide scale are most likely to succeed where the firms of the countries involved are advantaged in similar and complementary technologies. Given a more disaggregated sectoral classification of technological advantage, possible overlaps and complementarities between the firms of particular countries could be identified.

As well as indicating the scope for joint venture activity between large firms, the discussion of overlapping areas of specialisation between the firms of different countries may lead to a more general understanding of the conditions under which international technology transfer is likely to be successful. The technological accumulation approach suggests that the assimilation of foreign innovation is most easily accomplished where it is complementary to the existing pattern of technological specialisation of the country's firms. Attempts to establish entirely new fields of activity are rarely likely to be beneficial, except in so far as they are linked to the current activities of domestic firms (or multinational affiliates). While much of the recent literature on international technology transfer stresses various features of the market for technology, it may be more to the point to shift the emphasis of discussion towards the characteristics of the generation of technology within the firm.

This conclusion is especially pertinent when considering the paths of future innovation that offer the greatest potential to the largest firms of the major industrialised countries. It is these companies, most of which are multinationals, that now dominate the patenting statistics today. Such firms are particularly reliant upon innovative activity linked to fundamental research as a means of sustaining a high rate of growth, and they have often become embroiled in technological competition which has now moved to a world scale. The largest industrial firms are the most dependent upon research and scientific work, the fruits of which are more easily internalised and incorporated in technological advantage as represented by patents than is the case for production engineering. Smaller firms are more likely to be dependent upon the latter in their ability to innovate.

However, although each of the larger established firms may continue to steadily accumulate technology along specialised lines that are already reasonably clear, this does not mean that there is no scope for newer small firm entrants, especially in the areas of radical innovation in a newly emerging technological paradigm (mainly associated with microelectronics). It is therefore possible that a rise in the overall rate of innovation may be associated with a growth in activity on the part of such smaller firms, even if they rapidly become large themselves, or are taken over by larger firms. Even without new or smaller firms, established companies may broaden the extent of their technological specialisation. It has been shown that for Japan the degree of specialisation has broadened since the 1960s, despite the fact that the firms of most industrialised countries (11 out of 16) have become more specialised. What the technological accumulation approach suggests is that the firms of each industrialised country move out on a cumulative path in which the creation of new technological advantages depends on the pattern of advantages they have previously established. The existence of technological interrelatedness then allows for skills developed primarily in one sector to be utilised in another, and for a gradual movement between sectors. The broadening of specialisation is thus one of the possible forms of incremental change in the composition of innovation.

Whether countries should actively encourage a broader or narrower degree of technological specialisation amongst their firms depends *inter alia* on the relative size of the country concerned. Smaller industrialised countries (such as Sweden, Switzerland or Belgium) may wish to concentrate their efforts in areas in which their firms already have an established record. By contrast, the largest industrialised countries (such as the USA or Japan) may wish to maintain the broadest extent of technological specialisation possible. This has in fact tended to happen, as in his analysis of the variance of the RTA index, Pavitt (1987) demonstrated that the degree of technological specialisation is narrowest for the firms of the smaller industrialised countries.

The smaller industrialised countries may not trouble themselves too much about areas of decline, in which they are becoming increasingly dependent upon foreign firms whose research facilities are located abroad, providing this helps such countries to restructure their technological and productive activity more effectively in support of their innovative growth industries. Larger industrialised countries may be more inclined to try and avert any process of falling behind, particularly in what are regarded as important technological areas, perhaps with spinoffs elsewhere. This will be especially true the stronger is the relationship between the comparative advantage of a country's firms in productive activity and their specialisation in innovation. The form of the relationship between technological advantage and industrial growth is the subject of the remainder of the book, beginning with the growth of international trade and production in the 1950s, and the technological competition between US and European companies that it set in motion.

NOTES

1 Both measures rely on an estimation of moments of each distribution about its mean. If the nth moment of a distribution about the mean is given by μ_n, and $B_1 = \mu_3^2/\mu_2^3$ while $B_2 = \mu_4/\mu_2^2$, then the measure of skewness is approximated by

$$\gamma_1 = (\sqrt{B_1})(B_2 + 3)/2(5B_2 - 6B_1 - 9)$$

and the measure of kurtosis is

$$\gamma_2 = B_2 - 3.$$

2 In this case the mobility effect represents a shift towards industries that happen to have smaller rather than larger numbers of patents granted in total, while in the former case the mobility effect is associated with a move to 'larger' industries. Which industries are larger or smaller is clearly an arbitrary result of the industrial division that happens to be chosen, so the direction of the mobility effect has no economic significance.

3

A Dynamic Model of the Post-war Growth of International Economic Activity in Europe and the USA

3.1   Introduction

This chapter devises a framework for the analysis of technological competition between European and US firms since the 1950s, using the propositions of cumulativeness and incremental change examined in the previous chapter. It also draws upon and qualifies the Schumpeterian approach which has most often been used to analyse technological competition. Two rather broad but very important questions raised by the Schumpeterian tradition provide the context. The first has to do with generalised shifts in the competitive position of countries and their firms. In the nineteenth century, European countries pioneered many new technological developments, but by the turn of the century the USA had largely caught up. After the Second World War a new wave of innovations was led by US firms, and Europe (and Japan, followed by the newly industrialising countries) embarked on a process of catching up. Such long waves in industrial innovation, driven by technological leaders who pull a competitive fringe of imitators along behind them, are consistent with a Schumpeterian view of economic development. However, given that such a process is initiated, while Schumpeterian economists sometimes seem to suppose that the ensuing technological competition follows a fairly random course, the evidence of Chapter 2 suggests that it may follow more predictable channels.

Secondly, and related to this, a Schumpeterian model might suggest that the strength of 'catching up' in any industrial sector is determined essentially by the scope for imitation, as measured by the extent of the technological lead built up by the initial innovators. The theory of technological accumulation suggests instead that the rate of 'catching up' in a given sector is dependent more upon the potential for innovation amongst the firms of follower nations. That is, it is in the fields of their traditional strengths that European firms have the innovative capabilities to enable them to catch up and compete successfully with US (and today Japanese) firms. Following the findings of the previous chapter, it is reasonable to suppose that the rate at which European firms caught up depended upon the technological experience that they had previously accumulated in each industry.

Innovation is a cumulative process, such that the sectoral pattern of the technological advantages of a country's firms tends to remain fairly stable, irrespective of whether the overall rate of innovation is high or low (perhaps at different stages of a long wave). In a period of upswing the overall rate of innovation is high as firms of the follower countries catch up with the pace set by the leader. However, they catch up fastest (and overtake) in sectors in which they have a comparative advantage in innovative activity, and catch up slowest (or not at all) in sectors in which they are comparatively disadvantaged in their ability to innovate.

The model constructed here therefore relies on the existence of cumulativeness in the industrial pattern of technological change for the firms of any country, the evidence in favour of which has been considered in Chapter 2. Schumpeterian models of innovation usually emphasise discontinuity and 'creative destruction'. This may well be appropriate when addressing the question of why long waves in the overall rate of technological innovation come about (taking off towards the end of the last century, and in the immediate post-war period). However, questions connected with discontinuities in the overall rate of innovation are logically separable from questions connected with the sectoral pattern of innovation. The overall rate of innovation (as measured by the absolute level of patenting activity) appears to have risen in the 1950s and 1960s, and to have fallen in the 1970s and 1980s, but the sectoral pattern of this innovation continued to vary quite systematically across the firms of most countries.[1] The model developed here attempts to explore how the disruption of a major new wave of innovations (in this case, introduced by US firms in the early post-war years) affects the subsequent path of technological competition across sectors.

In order to provide a framework for the empirical investigation that follows, the model applies the idea of cumulative change in the pattern of MNC activity to the analysis of three issues examined in later chapters. The first issue is the ability of host country firms to respond to an increase in inward direct investment in their industries (the European response to the 'American Challenge', considered in Chapter 4). The second is the pattern of international economic activity of national groups of firms that evolves as a result of such competitive strengths or weaknesses (the subject of Chapters 5 and 6). Third are the conditions under which intra-industry production becomes important (the issue examined in Chapter 7). These three issues may be thought of as following in historical sequence in terms of the course taken by international technological competition in the post-war period. First, US firms invested substantially in European industries in the 1950s and 1960s; where they were able to European firms responded through export growth, and especially after 1970 with the expansion of their own international networks; and this has helped to lead to a steady increase in intra-industry production between Europe and the USA in certain sectors in the 1970s and 1980s.

Two main predictions are suggested by the framework of analysis developed in this chapter. Firstly, the European response to US firms is likely to have been greatest in the areas of their traditional technological strength (a view considered in Chapters 4 and 6). Secondly, intra-industry production is likely to grow in areas of mutual technological advantage (explored in Chapter 7). In each case, the cumulative development of MNCs is assumed to sustain in motion a process of competitive interaction between the firms of different countries.

The starting point is a criticism of the product cycle model (PCM), which represents an application of the Schumpeterian approach to economic development in the sphere of international trade. The PCM focuses on the discontinuities created by the emergence of entirely new products, and traces out their effect on international trade and the location of production as economies gradually return to a cost-determined equilibrium. However, product cycle ideas are of more limited value in a situation in which firms from more than one country engage in competing innovations (rather than simply catching up firms from the leading country). Where there are sectors in which the process of technological competition initiated by firms from the leading country results not only in imitation but in genuinely new innovations on the part of the firms of other countries, then a new model is required, embodying a more comprehensive treatment of innovation.

3.2   Innovation as a Dynamic Element in the International Economic Activity of Firms

The international economic activity of firms consists of three elements: trade, international production (financed by foreign direct investment, FDI), and non-equity resource transfers (as in licensing) across national boundaries. Until recently, the only dynamic theory of international economic activity widely discussed in the literature was the product life cycle, associated with Vernon (1966). It is argued here that the product cycle model (PCM) should be viewed as a special case in the development of a more general dynamic theory of international trade and production. However, it is a special case with particularly useful applications to the historical emergence of multinationals in manufacturing industry.

The earliest classical theories of trade (and, to the limited extent that it was considered, investment) were usually dynamic. Although Ricardo is generally thought of as the father of the modern doctrine of the static efficiency gains of trading in line with comparative costs, according to Thweatt (1976) he himself made little or no use of this doctrine, and Robinson (1974), Kregel (1977) and Maneschi (1983), amongst others, have emphasised the dynamic aspects of his approach to international trade. The classical economists essentially pioneered theories of the growth of production, with which the expansion of international trade and other economic links were viewed as complementary. Engaging in international trade was seen essentially as a means of promoting a more rapid rate of domestic economic growth. By contrast, neoclassical trade theories were reconstructed around a static model of exchange. Trade is to be welcomed as a means of improving domestic resource allocation at a given point in time. In this framework production fits awkwardly as an (end of period) stock rather than a flow concept, and the growth of production can only be artificially examined by the use of comparative statics (as suggested by Pasinetti, 1977). Technological innovation and the growth of output are in any case generally treated as exogenous in neoclassical trade models.

The PCM represented a partial return to the classical tradition, insofar as it dealt with innovation and the emergence of new products, as the basis for subsequent international trade and investment. Unfortunately, it suffered from two serious defects.

Firstly, the theory of innovation embodied in the PCM was inadequate, as it was viewed as an entirely market-determined process, a function of consumer demand and relative factor costs. The model's dynamism was thus provided by the growth of market demand for new products in the countries concerned. The theory of technological accumulation suggests that innovation is a firm-specific process, rather than a product-specific one. Firms, which are typically multiproduct, continually search for techno-logical improvements on their existing operations. New products and processes are not independent of one another, and they tend to fall into technological paths or trajectories followed by groups of competing firms and their suppliers and customers. Such trajec-tories are established in accordance with the new needs gradually created as production develops. On the whole, technical change is a continuous rather than a discontinuous process.

Since, in the PCM, the creation of new products was regarded as a series of discrete actions, the rules of neoclassical trade theory reassert themselves towards the end of each product cycle. The location of production is again determined by comparative costs once the product has become standardised. Indeed, the PCM is sometimes restated as a special case of the neoclassical approach, in which technology (and/or human capital) is treated as a factor of production, like labour or capital. In this scheme innovative countries export technology-intensive products, due to their rich 'endowment' of technology. In this version the dynamism of the PCM is lost, and it becomes a special case of the static theory of international trade and investment.

Secondly, and related to the first shortcoming, the PCM was constructed entirely at the level of the product, to the extent that the firm and even the industry were assimilated to the product.[2]

This becomes a major deficiency when examining the international-isation of industries, marked by a rise in specialisation in each location, as in the motor industry, and by the rise in intra-industry trade and production. Recent theories of international trade and investment have brought in the firm and the industry, but in doing so have often lost the dynamic theme which characterised the PCM. This is because the firm is not usually viewed as an initiator of production and innovation, but as a purely reactive agent within an established framework of market exchange (Casson, 1987). Each international market structure considered (with some combination of product differentiation and economies of scale) gives rise to a determinate solution for any representative firm (Helpman and Krugman, 1985). What is really needed is a model of an active firm, engaged in steadily expanding its international productive activity in oligopolistic markets (Cantwell, 1986a).

In Section 3.3 the PCM is re-examined. In Section 3.4 its application to the historical study of MNCs is investigated. In Section 3.5 a new approach to the international economic activity of firms based on cumulative technological change is developed, in which the PCM appears as a special case. Finally, in Section 3.6 the applicability of the model is considered in the light of some empirical evidence on the growth of US MNCs in European industries over the past 30 years.

3.3   The Product Cycle Model (PCM) Revisited

The PCM was developed by the Harvard school of economists,[3] and as such it provided background support for work on FDI by Knickerbocker (1973), Graham (1975 and 1978) and Franko (1976). It offered a useful explanation of US direct investment in Europe in the 1960s, and was also utilised by Graham and Franko as the basis for more elaborate accounts of both early and more recent European investment in the USA. However, the model considers only FDI of a trade-replacing kind, although it may ultimately lead to new trade flows in the reverse direction.

In the 1970s, faced by the growing internationalisation of industries, the basic model was amended to incorporate oligopolistic considerations. For this reason it is useful to look at two separate versions of the model, and here we follow the terminology of Buckley and Casson (1976) in distinguishing between the PCM Mark I and the PCM Mark II. An interpretation of the PCM is presented which is amenable to an adaptation in which technological competition between the firms of different countries is allowed for.

3.3.1   The Product Cycle Model (PCM) Mark I

The PCM Mark I was outlined by Vernon (1966) and was used implicitly in the work of Franko (1976). Innovations may generate intangible assets for certain firms, which give them 'ownership' advantages over other firms,[4] but in the PCM such innovations are always constrained by the markets specific to the country in which the firm originates. The creation of new products and processes depends on the existence of a large market to bear the high cost and risk involved in the supporting R&D expenditure. Thus, high incomes in the USA in the 1950s and 1960s led to the generation of particular kinds of ownership advantages on the part of US firms, such as in the production of a range of relatively expensive consumer durables by labour-saving processes. However, a technology cycle ensues with the growth of foreign incomes and hence markets, initially serviced by exports. The implication is that foreign competitors gain the potential of generating or acquiring similar ownership advantages.

For this reason, and not simply because of the more rapid growth of foreign markets, in the second stage of the 'maturing product' the relocation of production is considered. In this stage it is supposed that a more capital-intensive mass production begins to service a more price-elastic demand, so that imitators are more easily able to catch up. Also, the 'internalisation' advantages which the originating firms enjoyed, through the successful integration of research, production and marketing close to the site of innovating activities, begin to fade away. The growth of import-substituting FDI (in the region catching up) is spurred on by the erosion of ownership advantages by local rivals competing with exports from the country of innovation or threatening to enter the market.

However, it is important to note that the essential dynamism of the PCM is still provided by the evolution of market demand. It is the rise of incomes in the following countries that pulls along their firms and gives them the potential to catch up. Thus, the competitive position of foreign firms plays a secondary role in the PCM; in the model developed in Section 3.5 below this is elevated to a primary role.

The PCM Mark I is not very explicit on why international production rather than licensing (or non-equity resource transfer) displaces exports from the innovating country. International production prevails where innovating firms are thereby enabled to capture a full return on what remains of their technological lead, and where internalisation advantages may still be gained from the coordination of production and marketing. This happens where after-sales servicing is important in preserving market shares, or where capital goods need adapting to meet the requirements of local producers, though this goes beyond the remit of the PCM. However, the PCM Mark I is suggestive of the possibility that as the unique qualities of the innovator's intangible assets are diminished then licensing may become a more viable route of exploiting these ownership advantages abroad.

In the third and final 'standardised product' stage location advantages become paramount in determining the pattern of international trade and production. Such products are confronted by a highly price-elastic demand, giving rise to a more competitive market. Those firms remaining in the industry may try and move the emphasis of their activities downstream, replacing ownership advantages of a technological kind with ownership advantages grounded in marketing skills and distribution networks.

3.3.2   A Criticism of the PCM

The central difficulty with the PCM is its focus on the product, which becomes increasingly unsatisfactory as firms from various national origins interact and influence one another. In the PCM the international strategies of firms change solely in accordance with the maturation of the product. Changes in the international economic position of countries are brought in through introducing assumptions on patterns of demand and factor costs, and although these may provide a certain dynamism they constitute a driving force exogenous to the basic model. The firm comes to be equated with the product, being ideally at the outset a fairly new firm in a single business without international commitments. Even the industry is assimilated to the product through involving only those firms which produce it.

The PCM really runs out of steam after the location of production has begun to shift. Once it is allowed that local firms in lower-income countries are capable of their own innovation, perhaps drawing to some extent on licensed foreign technology but with their own distinctive characteristics, then an inherent dynamism is possible beyond some 'first stage' in the most developed country. As production grows in the follower country a new wave of innovative activity may begin, and there is absolutely no reason to believe that products and processes are progressively standardised. Walker (1979) provides an illustration of one such industry which does not conform to the product cycle prediction. When production of motor vehicles moved to Japan innovation in the industry increased rather than diminished, as Japanese firms were readily able to adapt their skills in electronic engineering. Baba (1987) takes this point further in the case of the colour television sector. Once television production had shifted to Japan not only was innovation resumed, but research continued at such a rate that it offset any trend towards standardisation. Japanese companies then relocated production in other industrialised countries such as the USA, rather than in lower-income countries; and the same has happened in the case of motor vehicles.

The PCM may be usefully thought of as an application of Schumpeter's (1934) model of innovation. Following Phillips (1971) and Freeman et al. (1982) this first model, which depicted the situation in the nineteenth century, is to be distinguished from Schumpeter's second (1943) model of innovation, which is said to correspond to the post-1918 period. In the earlier case innovation is believed to proceed in a science-push fashion, as entrepreneurial firms exploit the commercial opportunities arising from the work of scientists and private inventors. In this way new firms are established to pioneer entirely new products and industries. In the second Schumpeterian model innovation was regarded as a process partly endogenous to firms, reflecting the greater significance of R&D within large firms in the twentieth century. Yet even in this situation, Schumpeter's notion of creative destruction emphasises the role of major discontinuities rather than gradual adaptation in innovation. The implication is that, at least at the start of any cyclical upswing, new firms displace others.

Now the PCM is quite conveniently applied when only radical innovations introduced by new firms are considered. Its scope is decidedly more limited as continuous process innovations become increasingly important in inter-firm rivalry. Schumpeter recognised that the economic effect of innovation depended on whether the initiating firm was new to or already operated in the market, and on the existing market structure. During the twentieth century it became typical for innovation to be carried out by established firms, while markets became ever more concentrated. However, in the 1950s and 1960s when US FDI in Europe rose to its peak rate of expansion, new and existing US firms pioneered a variety of radical product innovations. The creation of a range of such major product innovations amounted to a wave of new technological development emanating from the USA and supported by the rapid growth of markets. The clear technological leadership of US firms and their ability to establish new areas of industrial activity (such as in computers or passenger jet aircraft) temporarily re-established the conditions to which the product cycle approach is best suited.

However, the PCM is ill-equipped to deal with the European or Japanese response to the American challenge, in which strong US, European and Japanese MNCs now compete side by side in the new international industries. The criticism of the PCM outlined here therefore reaches similar conclusions to that of Giddy (1978). Not only were European firms as capable of innovation (in many of the same industries) as their US counterparts. Once firms had generated a stream of substantial innovations of their own they became MNCs themselves or expanded their international production in other industrialised countries, and new corporate strategies emerged to suit the needs of a global organisation. Technological competition moved to a world arena, and firms reorganised their international networks of activities, increasing the degree of specialisation in each subsidiary which became less oriented to their own particular local markets.

3.3.3   The PCM Mark II

Since firms may benefit from the very multinationality of their activities (through the organisation of an improved international division of labour), Vernon (1979), has now recognised that there is a significant proportion of international involvement by US firms which the PCM is unable to explain. As early as 1971 Vernon recognised that the pure PCM was losing its explanatory capacity, and in 1974 he suggested a modified version in the PCM

Mark II by explicitly introducing oligopolistic considerations.[5] The Mark II model did not amend the PCM view of innovation, but it did allow firms to adopt non-innovative strategies in the later stages of the cycle. The emphasis moved away from the maintenance of ownership advantages of a technological kind, and towards barriers to entry due to scale economies, which enabled firms to preserve ownership advantages in the second stage of 'mature oligopoly'. Such advantages are attributable to the firm's size rather than its advanced technology.

In such a mature oligopoly, risk-minimising strategies on the part of the firm might better explain the displacement of trade by FDI than those associated with simple profit maximisation. Where a high ratio of fixed to total costs prevailed in industries characterised by economies of scale, security became a more important consideration for the firm relative to profitability. Vernon argued that this led to cross-investment (that is, intra-industry production) to reduce the threat of subsidiary price cutting in the domestic market of each large firm.

In this second model international production is established not only as a means of exploiting ownership advantages by producing in favourable locations, but it also plays the role of preserving equilibrium in the international market. In this regard the PCM Mark II picks up one specific aspect of the shift in emphasis for large firms from a particular market abroad to the world market. However, the Mark II version retains many of the deficiencies of Mark I. It still subordinates the international strategies of firms to the maturity of their industry (which is also true of the alternative industry technology cycle of Magee, 1977). In the PCM Mark II, it is where the industry is mature that strong firms from various countries are more likely to coexist, and the structure of costs impels them to be more responsive to each other's moves. In fact, MNCs often compete strongly through innovation in industries which are far from mature (Chapter 7 below examines the possibility that cross-investments may result from attempts by each group of firms to gain access to complementary foreign innovation), while elsewhere intra-industry trade and production is more widely accounted for by the expansion of globally integrated MNCs than by an oligopolistic exchange of threats. The Mark II amendments are quite inadequate in coming to terms with the major changes in the international economic activity of manufacturing firms since the 1960s.

3.4   The Explanation of International Production by Innovative Manufacturing Firms before 1914

The PCM fits a Schumpeterian model of innovation designed to help explain industrial development in the nineteenth century. An entrepreneurial firm pioneers a new product or process and creates a new market for itself partly by taking the markets for substitutable products. Imitative followers then gradually whittle away the high profits of the innovative leader. The entrepreneur may move on to fresh innovations elsewhere, but in the industry in question the originator firm faces a continual decline in its market share, at a rate determined by the capability of its competitors to eliminate its technological lead. This is very much like the PCM Mark I in which the survival of the innovator in all phases of the cycle depends upon the continued existence of its technological ownership advantages.

Although the PCM was originally devised as an explanation of US trade and investment in the 1950s and 1960s, it is, like the analogous Schumpeterian model, more directly applicable to a much earlier period of innovation around the turn of the century. It fits the case of European FDI in US manufacturing industries before 1914 quite well. Here Sanna Randaccio (1980) reaches similar conclusions to those of Buckley and Roberts (1982) and Franko (1976). One factor stands out above all others in distinguishing successful from unsuccessful European investments. The investing firms all pioneered comparatively new products or processes, but the successful firms were able to preserve their ownership advantages (usually of a technological kind) due to the absence of strong US competitors in the same industry. Where this condition was met, US trade barriers often played their part in precipitating a switch from trade to direct production in the USA in the product cycle framework.

Three European firms held especially strong technological leads: Courtaulds had its rayon patents, Bayer led in organic chemicals (at a time when US dye firms were mainly assembly plants working on German intermediate products), and the Belgian company Solvay had a unique material-saving method of soda manufacture. The US subsidiaries of these European MNCs were unchallenged by local competitors before 1914, and their technological ownership advantages were preserved.

By contrast, the Anglo-Swiss Condensed Milk Company, and

Daimler and Fiat in the motor vehicle industry, were firms which despite an innovative lead were soon forced to close US subsidiaries prior to 1914, as US rivals quickly caught up. Perhaps an even more telling example of such a forced divestment is that of the German electrical company Siemens. Siemens operated a general range of activity in the electrical industry in which smaller, more specialised European firms were able to establish successful US subsidiaries, in branches where US competitors had not yet developed. However, even Marconi, for instance, had lost its position by 1919.

The period around the turn of the century and that which followed 1945 both witnessed major innovations (for example, aluminium, the motor vehicle and rayon before the First World War, and the computer, synthetics and jet aircraft after the Second World War). Each of these innovations gave rise to a new industry, having endowed particular firms with the technological ownership advantages required by the PCM for a sustained growth in international trade and investment. However, each of these periods of technological leadership was followed by a phase of technological convergence associated with the emergence of more settled industry structures internationally. This suggests, therefore, that the applicability of the PCM is limited to a quite specific set of historical conditions.

3.5 An Alternative Model of Technological Competition between US and European Firms

A viable model of technological competition in a modern international industry must deal with the competitive interaction between firms pursuing active strategies, and to do so it can draw on the theory of technological accumulation as an underlying driving force. The explanatory power of the PCM as a conceptual apparatus is at its greatest where major innovations create new industries. Yet this is an exceptional situation. The restructuring of industries more usually proceeds through the gradual adaptation of production by innovative firms. The previous technological experience of such firms, and the problem-solving activity in which they are currently engaged, set the scene for the course of future innovation. Until there is a new stream of breakthroughs based on a different set of fundamental discoveries, firms at the existing frontier of progress tend to establish dynamic advantages over others. If so, it is likely to have been in those sectors in which European technological efforts were historically concentrated that European firms would have most readily been able to respond to the American challenge of the 1950s and 1960s.

The comparative strengths of firms affect their international strategies. Isolating the USA and Western Europe for the purposes of the present discussion, the pattern of technological advantage on either side of the Atlantic helps to establish the likely industrial distribution of thriving firms in each region. It is useful to think of three possible groups of industries: (i) those in which the leading US firms are strong, while European firms are weak and more narrowly specialised; (ii) those in which US firms are weak and European firms strong; and (iii) those in which strong US and European firms coexist in competitive interaction. For the sake of simplicity differences in technological advantage between the firms of different European countries are disregarded for now, though it is easy to extend the analysis to cover national diversity within Europe.

In the first two situations strong firms will expand internally rather than licensing other firms. Their choice between exports and international production in servicing the foreign market will be largely determined by the location advantages of producing in the USA and Europe, including allowance for trade barriers and transport costs, and the relative size of markets. However, it is likely that at least some international production will prevail, so that firms from the advantaged region are able to make the gains which follow from a global network of operations, and thus build a more secure position for themselves in the international industry.

By contrast, technologically 'weaker' firms in these industries will be far more specialised, and their international operations much smaller. To the extent that they diversify their activities, they will be far more prone to enter into joint ventures, or to license new technological know-how to foreign firms. However, this needs qualification to the extent that one region begins with a generalised technological leadership. Considering case (ii) above, in the early post-war years US firms in all industries may have been boosted by a new spirit of technological competition, sometimes entering into entirely new activities and developing new skills. Meanwhile, European firms were emerging from a period of cartel and similar agreements, and a lack of urgency in their technological activities. In this scenario, US firms were sufficiently confident to pass down the product cycle route, of exporting to and then producing in Europe. In this case, the new wave of competition disturbed the lethargic Europeans, until they drew on the hidden reserves of their own technological expertise to restore their position.

The evidence of the previous chapter has shown that it is reasonable to suppose that as a general rule the firms of any industrialised country hold technological advantages in particular industries for rather long periods of time. From this it would follow that European 'catching up' represented the revitalisation of traditionally strong sectors (or, at least, the underlying types of technological activity which support them), rather than the relocation of new products and industries as such.

In this context, the most interesting possibility is that described by (iii) above, in which firms of differing nationalities have specific mobile skills of their own in similar technological areas. In this situation, intra-industry trade and production (between the USA and Europe) is likely to develop, together with international cross-licensing agreements between major competitors. For each of the firms as a whole rather than for particular products or technologies, licensing and joint venture cooperation is complementary to rather than substitutable for international production. Under these conditions, intra-European cooperation and joint ventures, in which each partner has a clear contribution to make, are more likely to be successful. Indeed, in these industries European 'catching up' may often have relied in part initially on the licensing of new US technology. The case of industries (ii) and (iii) amounts to a 'virtuous circle' (using the notion of cumulative causation as expounded in this context by Cantwell, 1987, and Dunning and Cantwell, 1989), in which inward investment acts as a beneficial competitive stimulus.

By dividing industries into three groups in this way, three outcomes of the interaction between the European operations of US firms in the 1960s, and the technological advances and international involvement of European firms can be distinguished. In the first group, US firms have retained technological leadership which supports a strong position in foreign affiliate sales and/or exports. Their European entry may have acted to compound a 'vicious circle' in the activity of European firms, by challenging them more effectively in their domestic markets. In the second group, US firms entering Europe had little long-term advantage over European competitors, but were mainly drawn abroad by locational attractions in the early days of the EEC and have since retreated. In the third group European firms caught up and recovered technological advantage, and a network of intra-industry trade and production has often resulted. In all three groups, as the strongest firms broaden their experience of economic activity through a wider or renewed range of technological advantage, they more readily diversify industrially and enter international production. In the course of doing so they become global MNCs, and they reorganise their international operations accordingly.

3.5.1   The Model

The procedure followed is to examine the stages of development that characterise each of three divergent industries, following the establishment of an innovative lead amongst the firms of one particular region. The two regions considered, the USA and Europe, are denoted by U and E, and in each of them there are three industries (1, 2 and 3) corresponding to cases (i), (ii) and (iii) above. The indigenous firms of each industry in U and E compete at home with imports from the other region. It is assumed that initially the growth of each industry is equalised across countries, and is the same for each of its component parts, so that the market shares of domestic firms and imports are constant. In industry 1 US firms hold strong ownership advantages and have hence acquired relatively high market shares, in industry 2 European firms have strong ownership advantages, while in industry 3 a more balanced situation prevails.

Now for U, let U_i be the total domestic market for industry i ($i = 1, 2, 3$), and let U_{ij} be that part of the market serviced by the local sales of indigenous firms. Let U_{im} be that part of the market serviced by imports from abroad. Likewise, E_i is defined as the total sales in industry i in E, E_{ij} represents the local sales of European firms, and E_{im} designates imports. The following relationship now hold:

$$U_i = U_{ij} + U_{im}$$
$$E_i = E_{ij} + E_{im}$$

It is supposed that in the first instance in each sector the proportion of innovation accounted for by the firms of each region is positively related to their proportion of total production in the regions taken together, such that their proportional growth and market shares are given and equal:

$$\dot{U}_i = \dot{U}_{ij} = \dot{U}_{im} = \dot{E}_i = \dot{E}_{ij} = \dot{E}_{im}$$

where dots over the letters indicate proportional rates of growth, e.g.:

$$\dot{U}_i = (1/U_i)\,(dU_i/dt)$$

Under these initial conditions the innovative activity of each firm continues in proportion to its existing ownership advantages. Ownership advantages are defined here as advantages that lower the unit costs of a firm relative to its major rivals in its industry. The stronger are ownership advantages (the higher is the share of patenting activity) the larger the market share held by the firm. Since U's firms have relatively stronger ownership advantages in industry 1 than in industry 2, their market shares are higher. In the terminology of Chapter 2, US firms have RTA values that are greater than 1 in industry 1, less than 1 in industry 2, and around 1 in industry 3, and this is reflected in their existing market shares. If ownership advantages improve (the share of patenting in the sector rises) then the firms concerned increase their market share. This is what happens in the first stage of the sequence due to a wave of major new innovations in U. The first stage may be called 'the Schumpeterian technological leadership phase'. Indigenous firms producing in the USA now enjoy the benefits of an increased rate of innovation at home, eroding the share of imports. Thus, it now follows that:

$$\dot{U}_{ij} > \dot{U}_i \quad \text{and} \quad \dot{U}_i > \dot{E}_i$$

In the second stage, firms based in U begin to exploit their new innovative strength abroad through exports. The second stage is the 'US export growth phase'. This gives:

$$\dot{E}_{im} > \dot{E}_i$$

The changes of the first and second stages also result in a rise in (U_{ij}/U_{im}) relative to (E_{ij}/E_{im}) in each industry, reflecting the improved ownership advantages of U's firms. In the third stage, European recovery begins, while US firms establish European subsidiaries in response, the sales of the latter being denoted by E_{is}. This is the 'US firm international production growth phase'. Hence:

$$\dot{E}_i > \dot{U}_i \quad \text{and} \quad \dot{E}_{is} > \dot{E}_i, \text{ where}$$
$$E_i = E_{ij} + E_{im} + E_{is}$$

So far, the product cycle pattern has been adhered to, but in the fourth stage the path of each industry diverges. The fourth stage may be thought of as 'the early European recovery phase'. In industry 1 some European firms are driven out of the market E_1, and are taken over by US firms (so E_{1s} increases by means of acquisition). The European firms find it difficult to meet the challenge, although the remaining firms selling E_{1j} with the aid of their own technological efforts, also begin to benefit from licensing technology from U. Those sales in industry E_i or U_i which are directly attributable to the licensing of foreign technology are referred to as E_{il} and U_{il}. Thus:

$$E_i = E_{ij} + E_{im} + E_{is} + E_{il}$$

In industry 1 the share of foreign-owned affiliates still continues to rise $(\dot{E}_{1s} > \dot{E}_1)$.

In the fourth stage in industries 2 and 3, U's comparatively weaker firms begin to license E's comparatively stronger firms, and European exports to the USA begin to grow again. In industry 2 the growth of independent and licensed firms outstrips that of foreign-owned firms $(\dot{E}_{2j} > \dot{E}_{2s}, \dot{E}_{2l} > \dot{E}_{2s})$, while import penetration in the USA rises $(\dot{U}_{2m} > \dot{U}_2)$. In industry 3, while US-owned affiliate growth in Europe continues to be strong, European firms have begun a successful export drive of their own $(\dot{U}_{3m} > \dot{U}_3)$. However, in industry 3 it need not be the case that the domestic sales of European firms (whether due to licensing or not) begin to grow faster than those of foreign-owned subsidiaries.

In the fifth stage industry 1 settles back to a position in which shares of production and innovation are again equal, firms grow at the same proportional rate, and market shares are stabilised. However, this only happens once $(E_{1m} + E_{1s})$ has risen considerably relative to E_{1j}. More generally, though, the fifth stage constitutes 'the European firm international production growth phase'.

In the fifth stage in industry 2 foreign-owned affiliates are taken over by the stronger European firms, which establish their own subsidiaries in the USA, and licensing by European firms becomes less important. This is represented by:

$$E_{2s} = 0, \dot{E}_{2j} > \dot{E}_{2m} > \dot{E}_{2l}, \quad \text{and} \quad \dot{U}_{2s} > \dot{U}_2$$

In industry 3 the stronger European firms establish their own subsidiaries in U ($\dot{U}_{3s} > \dot{U}_s$), and a network of cross-licensing arrangements develops between each of the firms ($\dot{U}_{3l} > 0$ just as $\dot{E}_{3l} > 0$).

The sixth stage represents an 'intra-industry production phase', though this is essentially a characteristic of industry 3. In industry 2 European-owned subsidiaries in the USA take over comparatively weaker US firms, while the comparatively weaker European firms begin to license the remaining US firms ($\dot{U}_{2l} > \dot{U}_{2j}$).

In industry 3 foreign subsidiaries take over the relatively weaker domestic firms in each region, as each of the major firms begins to establish a global network of activities ($\dot{E}_{3s} > E_3$ and $\dot{U}_{3s} > U_3$).

The seventh stage may be described as 'the balanced innovation phase', in which the early technological leadership has been completely eroded. In the seventh stage all three industries settle back down to a path of equal proportional rates of growth across groups of firms and stable market shares. In industries 1 and 2 the overall world share of the advantaged groups (from U and E respectively) has increased through the establishment of international production, while in industry 3 intra-industry production has come into being. The final position is:

$$U_i = U_{ij} + U_{im} + U_{is} + U_{il} \quad U_{1s} = U_{1l} = 0$$
$$E_i = E_{ij} + E_{im} + E_{is} + E_{il} \quad E_{2s} = E_{2l} = 0$$

Of the three possible industries described here, the product cycle model most closely resembles the course followed by industry 2. However, there are at least two crucial differences. Firstly, the

PCM begins and finishes at a position of static rather than dynamic equilibrium, since it deals essentially only with the major innovative breakthroughs which cause the disturbance, while it disregards the continuous process of adaptive innovation. Secondly, in the PCM it may be envisaged that the exports and international production of European firms grow as they imitate the US leaders, rather than due to their inherent technological capability partly inherited from their past experience. Moreover, in the PCM the industry would not be expected to settle here, but production would be expected to begin to shift from Europe (and the USA) to, say, the newly industrialising countries. In this account this aspect of the PCM has been ignored, since it is assembly types of activity that have generally shifted to developing countries, rather than the research-intensive activities that lie at the heart of technological competition, and which have tended to remain within the industrialised countries.

The relevance of the product cycle idea is really restricted to the first three stages of the model, and it helps to describe only one phase in a more general approach to technological competition. In industries in which innovation is an ongoing phenomenon, the pattern of technological advantage across countries and firms is more important in determining the course of expansion of international economic activity, than the gradual maturation of particular products.

Two aspects of this discussion are crucial. Firstly, it has been suggested that the European response to the rapid growth of US MNCs depended upon their underlying technological capabilities (the pattern of their comparative technological advantages). Secondly, as an extension of this line of reasoning, intra-industry production in any sector tends to become established and to grow between countries whose firms are both relatively advantaged in the industry in question. These propositions are examined in the chapters that follow.

3.6   The Application of the Model

The model outlined in Section 3.5 concentrates on just one aspect of recent developments, that is the evolution of technological competition between US and European firms, and its variation across industries. Although the model will help to inform the subsequent discussion, one particularly important consideration has been omitted up until now for the sake of simplicity. That is, the model has abstracted from the continuing trend towards the greater internationalisation of all industries. Firms increasingly gain advantages from multinationality *per se* and the development of a stronger division of labour across a globally integrated network of activities, rather than the exploitation of any particular set of technological or similar advantages. Moreover, the establishment of a network of international production and research has become a means of strengthening technological accumulation and competitiveness at the level of the firm.

For this reason, the claim of the model that the international production of US MNCs in industry (ii) fades away, while European MNCs never establish US production facilities in industry (i), is too strong. Even though they lack the greater technological capacity of their major foreign rivals, MNCs may well become established and survive through the closer integration of a range of complementary activities carried out in different countries. Firms that are directly represented in all their major markets are able to seize upon every opportunity for growth more rapidly.

These points should be borne in mind when assessing the applicability of the model as an explanation of changing patterns of international trade and production. This section concludes with some preliminary impressions on which industries fall within the groups (i), (ii) and (iii) described above. The industry classification differs depending on which European country is under discussion; the pattern of the technological strengths and ownership advantages of firms varies between European countries, just as it does between Europe and the USA.

Trends in the share of the international production of US MNCs in the industrial output of the four largest European economies between 1957 and 1977 are depicted in Table 3.1. For each industry this measures what can be defined using the notation set out above as $E_{is}/(E_i - E_{im})$. Although it would be helpful to distinguish between industries at a fairly low level of disaggregation, the data do not allow for this. Taking the 21 sectors on which data are available on the changing share of US affiliates between 1957 and 1966 (six in both West Germany and the UK, five in France, and four in Italy), the international production of US-owned firms

Table 3.1 The share of European industrial output accounted for by US majority-owned foreign affiliates, 1957–1977 (%)

	West Germany			UK			Italy			France		
	1957	1966	1977	1957	1966	1977	1957	1966	1977	1957	1966	1977
1. Food Products	1.92	3.65	8.08	4.93	6.26	8.62	N.A.	3.11	4.49	N.A.	2.48	3.33
2. Chemicals	1.17	5.06	7.74	11.08	16.62	16.05	3.98	7.78	9.89	3.70	8.08	12.56
3. Metals	0.39	1.29	3.74	2.13	5.38	4.80	0.86	3.75	1.71	0.70	1.64	1.64
4. Mechanical Engineering	3.43	8.83	12.20	8.77	15.57	22.05	3.30	14.07	13.59	2.11	7.46	16.66
5. Electrical Equipment	1.71	}13.93	5.11	7.10	}14.96	13.03	4.13	16.74	9.54	3.66	}9.28	7.86
6. Transport Equipment	11.49		12.08	9.45		26.64	N.A.	1.88	2.59	3.72		10.62
7. Other Manufacturing	0.52	1.88	6.99	2.62	4.15	8.09	N.A.	1.78	2.64	0.74	2.25	3.99
Total Manufacturing	2.06	5.29	7.81	5.23	9.01	12.34	1.44	4.74	5.10	1.42	4.49	6.93

N.A. = Not available

Source: US Department of Commerce, *US Business Investments in Foreign Countries* (1960), *Survey of Current Business* (August 1974), *US Direct Investment Abroad, 1977* (1980); UN, *Yearbook of Industrial Statistics* (various issues).

increased its share of domestic output in all 21. This is the period of US expansion in Europe across all industries, as suggested by the model.

US majority-owned affiliates saw their share of manufacturing output rise from 2% to 5% in West Germany, from 5% to 9% in the UK, and from about 1.5% to 4.5% in France and Italy. Considering these four European economies jointly, the total manufacturing output of US foreign affiliates expanded at an annual growth rate of 14.1% between 1957 and 1966, whereas total industrial output grew at an annual rate of 4.8%. That is, US affiliates grew roughly three times as fast as the average of all industrial firms considered together. In the later period growth measured in nominal terms rose, because of higher rate of inflation. From 1966 to 1977 the annual rate of growth of US manufacturing affiliates in the four largest European economies was 14.5%, while total industrial output rose at 11.7% per annum. That is, the comparatively faster rate of growth of US firms' international production slipped from three to only one and a quarter times the average. This illustrates the process of European catching up under way in a number of sectors.

In the 25 country-sector combinations for which data are available (electrical and transport equipment must be considered jointly for West Germany, France and the UK), the share of US affiliates actually declined in six, and remained unchanged in a seventh, between 1966 and 1977. This is despite the continuing internationalisation of industries referred to at the beginning of this section. The sectors in which European revival was most marked in the post-1966 period were transport equipment (particularly motor vehicles) in West Germany, and perhaps electrical equipment; chemicals and metals in the UK; metals, mechanical engineering and electrical equipment in Italy; and metals in France. In these cases it seems likely that there are at least some sub-sectors that are representative of the pattern depicted in the model by industries (ii) and (iii). A more detailed empirical assessment of the European response to the 'American Challenge' is to be found in the next chapter.

NOTES

1 A further discussion of the relationship between the two can be found in Pavitt and Soete (1982).
2 The model was not applied simply at the product level, but was used as a general explanation of the rise of US direct investment in post-war Europe, in those industries in which US firms had an innovative lead.
3 The first complete exposition of the theory is usually credited to Vernon (1966), but see also Vernon (1971 and 1974), Hirsch (1967), Wells (1972), and the associated work of Hufbauer (1965 and 1970).
4 Here Dunning's eclectic paradigm is called upon. In its dynamic formulation, ownership and internalisation advantages are developed and exploited by firms in the course of their growth, in the context of changing location advantages associated with production in particular countries or regions. For such an approach in cross-section analysis, see Dunning (1982). The eclectic paradigm is discussed further in Chapter 9.
5 See Vernon (1971) and Vernon (1974) respectively.

4

The Evolution of Technological Competition between US and European Firms

4.1 Introduction

Early studies of the post-war rise in US investment in Europe, being primarily concerned with the technology 'gap' between US and European firms, typically adopted the theoretical framework of the product cycle. However, one such study (Dunning, 1971) went beyond this, and suggested that the rapid growth of US foreign affiliates not only improved the position of European locations as a base for the production and export of research-intensive goods, but that in some sectors it acted as a competitive spur to indigenous European firms and technological capacity. Similar conclusions are to be found in related work by the same author at around this time (see, for example, Dunning, 1970a and 1970b). What has been termed here a 'virtuous circle' was described by Dunning at that time as a 'technological multiplier' effect.

This chapter extends the statistical analysis of Dunning's 1971 paper[1] to take account of the European response between the mid-1960s and mid-1970s, while dropping the product cycle approach and placing greater emphasis on the identification of those sectors in which European firms proved most capable of recapturing lost ground. The chapter describes in some detail the sectoral variations in the European response to the earlier growth of US firms. Drawing on the conceptual framework developed in Chapter 3, it investigates the evolution of technological competition between firms amidst an environment in which European industry was catching up, with growth rates reaching higher levels in Europe than in the USA. As interest lies in the process of catching up rather than in the establishment of US technological leadership that preceded it, stages three to six of the model of the previous chapter are the relevant ones.

The focus is on how well Europe caught up in terms of exports rather than production of manufactured goods. In other words, the emphasis is on shifts in the competitive advantages of alternative locations for the servicing of international markets, rather than on the rate of growth of local markets in their own right as in the PCM. The improvement of European export shares is then related to the expansion of exports by US-owned foreign affiliates in Europe, to see the extent to which, at least in the early stages of catching up, US firms continued to grow relative to European firms (the extent of the 'American Challenge'). The analysis is then extended to look at the response of indigenous firms to the challenge, and at how this varied across sectors, distinguishing between the three types of industry suggested in Chapter 3.

Hymer and Rowthorn (1970) had found that as a general rule the Americans felt challenged themselves as the technological gap narrowed, in spite of the fact that for most of the 1960s the European subsidiaries of US firms were normally growing faster than non-US firms in Europe. Selective industry studies (for example those of Lake, 1976a and 1976b, on the UK pharmaceutical and semiconductor industries) have shown that, in certain cases, the direct presence of US affiliates had exerted considerable competitive pressure on host country firms to undertake R&D, which led to an improvement in the competitiveness and growth of indigenous companies. In such sectors technological rivalry between US firms and host country firms accelerated the rate of local innovation. In those countries and sectors in which indigenous firms were well placed to respond, the rate of technological accumulation of European firms increased. As this happened, local firms began to catch up with the rapid rate of growth of production and exports achieved by the US-owned foreign affiliates operating in Europe.

Section 4.2 deals with the catching up of European locations as measured by their share of world trade in manufactures. The role of US foreign affiliates in promoting this process, and the reaction of indigenous firms to the presence of tehnologically advanced foreign-owned companies is the subject of Section 4.3. Finally, some evidence on the significance and the direction of non-affiliate licensing as the technological competitiveness of firms altered is considered in Section 4.4. The analysis refers to the period 1955–75. By the mid-1970s, it is inadequate to think of the catching up of European firms (as opposed to domestic industries) in terms of European export shares, since they had by then begun to increase their international production faster than their home country exports. The situation since the early or mid-1970s therefore requires a study of overall world market shares in terms of the total international economic activity of firms, but this is left for Chapter 5.

4.2 *European Catching Up, and the Relative Growth of European and US Exports of Manufactures, 1955–1975*

The growth of European and US exports for a range of manufacturing industries between 1955 and 1975 is documented in Table 4.1. A distinction is drawn between the original six founder members of the EEC and the rest of Western Europe. The data collected show that from 1955 to 1975 the percentage of European plus US exports accounted for by European countries rose from 67.9% to 78.2%, with the largest part of the increase coming in the first decade and with EEC countries scoring more significant increases than other European countries.[7] For this period as a whole, Europe's competitive position improved most in pharmaceuticals, motor vehicles, rubber products, and coal and petroleum products. Only in one sector, namely professional and scientific instruments, did the European share of European plus US exports actually decline, though, in the second half of the period, some of the earlier loss of markets by the USA in the aircraft, office machinery and motor vehicles industries were recouped.

Within Europe, export growth rates (in value terms) varied widely, but only the UK performed less well than the USA for both periods, as shown in Table 4.2. Italy, West Germany and the Netherlands achieved the highest growth for the 20 years, although, in the second decade, France joined this group. Catching up seems to have been fastest and most effective in the largest

Table 4.1 The growth in exports of selected products: USA and Europe, 1955–1975

Product	USA			EEC SIX			EUROPE		
	1955–1965 (1955=100)	1965–1975 (1965=100)	1955–1975 (1955=100)	1955–1965 (1955=100)	1965–1975 (1965=100)	1955–1975 (1955=100)	1955–1965 (1955=100)	1965–1975 (1965=100)	1955–1975 (1955=100)
Food and Drink	134.2	392.5	526.7	265.7	485.8	1290.8	246.7	432.6	1067.2
Tobacco Products	196.3	327.3	642.5	592.8	573.6	3400.3	231.2	469.3	1085.0
Chemicals, n.e.s.	240.3	366.7	881.2	305.4	559.3	1708.1	273.6	546.2	1494.4
Pharmaceuticals	112.4	342.6	385.1	315.1	518.7	1634.4	259.8	516.8	1342.6
Ferrous Metals	93.3	394.8	368.3	214.9	487.0	1027.2	209.6	463.0	970.4
Non-ferrous Metals	268.8	254.0	682.8	257.8	310.9	801.5	249.8	336.2	839.8
Fabricated Metal Products	179.8	344.9	620.1	246.4	511.0	1259.1	216.5	513.7	1112.2
Mechanical Engineering, n.e.s.	193.2	414.0	799.8	331.9	566.9	1881.5	272.1	522.6	1422.0
Agricultural Machinery	517.4	332.1	1718.3	701.7	540.5	3792.7	652.5	432.9	2824.7
Office Machinery	436.6	568.4	2481.6	577.3	516.4	2981.2	464.8	560.5	2605.2
Electrical Equipment	215.0	467.8	1005.8	366.5	537.9	1971.4	301.6	545.8	1646.1
Motor Vehicles	146.0	510.7	745.6	435.2	555.5	2417.5	342.9	497.0	1704.2
Aircraft	150.2	564.5	847.9	2169.4	330.3	7165.5	522.9	345.6	1807.1
Textiles and Clothing	129.5	296.0	383.3	206.0	381.0	784.9	178.2	391.4	697.5
Rubber Products	128.2	392.3	502.9	262.6	608.5	1597.9	282.4	589.7	1655.3
Non-metallic Mineral Products	141.8	277.7	393.8	238.4	452.3	1078.3	198.9	468.3	931.4
Coal and Petroleum Products	71.4	222.4	158.8	194.3	659.0	1280.4	187.1	681.4	1274.9
Professional Instruments	497.4	445.0	2213.4	292.8	462.5	1354.2	310.7	468.9	1456.9
Paper and Allied Products	197.2	373.3	736.9	295.0	569.2	1679.1	232.4	601.1	1397.0
Printing and Publishing	240.7	246.6	593.6	351.0	428.3	1053.3	326.7	446.8	1459.7
Lumber and Wood Products	154.9	611.9	947.8	338.5	649.3	2197.9	243.9	669.9	1633.9
Total Manufacturing	177.1	407.7	722.0	284.5	510.1	1451.2	251.2	488.3	1226.6

Source: United Nations, *Commodity Trade Statistics*, 1955, 1965 and 1975.

Table 4.2 The growth in exports of European countries and the
USA, 1955–1975

Country	1955–1965 (1955=100)	1965–1975 (1965=100)	1955–1975 (1955=100)
USA	177.1	407.7	722.0
EEC six:	284.5	510.1	1451.2
West Germany	315.9	514.5	1625.3
France	228.8	522.2	1194.8
Italy	359.9	490.5	1765.3
Netherlands	276.4	538.7	1489.0
Belgium and Luxembourg	248.5	472.1	1173.2
UK	171.4	329.6	564.9
Denmark	328.8	389.4	1280.3
Other Europe	208.7	447.4	933.7
Europe	251.2	488.3	1226.6
Total	228.1	468.2	1068.0

Source: United Nations, *Commodity Trade Statistics*, 1955, 1965 and 1975.

countries of the original EEC six. Rates of growth were generally higher in the 1965–75 period, but this was at least partly a consequence of inflation. Overall, it seems to be the case that in the period since the mid-1960s, for manufacturing as a whole, some convergence occurred between the rates of export growth of European countries and the USA. The catching up of European countries as production sites from which to serve world markets was at its strongest in the late 1950s and early 1960s.

4.3 The Role of US International Production in European Catching Up, and the Response of Indigenous European Firms

The next issue is the extent to which the catching up of European locations can be attributed to the international production of US firms in Europe. For the reasons discussed in the previous chapter,[3] the increasing attractions of a production outlet in Europe resulted in a continually rising ratio between the output of US affiliates and exports from the USA (the E_{is}/E_{im} ratio using the terminology of

Chapter 3); the increase being much the same for both the 1957–66 and 1966–75 periods. Table 4.3 sets out the figures. The faster growth of European markets and the re-emerging competitive potential of European companies were reinforced by the formation of the EEC. The beginnings of the rationalisation of production of US firms in Europe further contributed to the trend towards the replacement of exports to European markets by international production. The catching up of Europe and its improvement in technological competitiveness meant that the US trade position deteriorated, but in a system of pegged exchange rates the US dollar became over-valued such that US firms could buy up European companies more cheaply, which became a secondary factor in the rise of US FDI in Europe (though one which has been emphasised by Aliber, 1970).

Table 4.3 shows that Italy, Belgium, Luxembourg and the Netherlands recorded the largest increase in the E_{is}/E_{im} ratio, thus indicating the increasing locational advantages of producing in the smaller European countries in the EEC. By contrast, the E_{is}/E_{im} ratio of the UK hardly rose at all, which in part reflects the longer standing of the UK as a host to the investments of US MNCs (as described by Dunning, 1958). Among industrial sectors, the increased propensity to favour a European location was common to all except electrical equipment. It is notable that in the chemicals and rubber product sectors in which European firms have been technologically strong historically the rise in the E_{is}/E_{im} ratio was above average, while in the electrical equipment industry in which they have been weak relative to US firms there was a slight fall rather than a rise in the ratio. Above-average increases were also recorded in the paper and metals sectors.

Although the rise in the E_{is}/E_{im} ratio reflects the increasing attractiveness of European locations in the 1955–75 period, it would be wrong to think simply in terms of a direct substitution between US exports and local European production. Local sales of US affiliates may include imports from their parent companies, while part of these sales may be exported (though perhaps to the same third country markets served by US exports). In the case of rationalised investments, it is less likely that the exports of affiliates may be thought of as a substitute for US exports than in the case of import substituting investments. And even in the latter case, as various surveys have shown,[4] exports and sales of foreign subsidiaries are often complementary with one another, for example the latter may help to create a demand for goods from the parent company which it is not economic to produce locally.

US firms contributed to the growth of European exports both directly and indirectly, through the licensing of some European firms and the competitive stimulus that they provided for others, as well as being affected by it (being attracted to Europe as a prospective platform for exports). In so far as the growth of exports depended upon the growth of US international production, it seems that they become more responsive in the later period. Simple regressions performed using observations on local exports and US inward direct investment across the eight European countries or groups of countries[5] yield the following results:

For 1955–65

$$\Delta X = 201.3 + 0.33 \, \Delta K_{US}$$
$$(2.1)^*$$
$$R^2 = 0.42 \qquad \text{Standard error/mean} = 0.19$$

and for 1965–75

$$\Delta X = 295.5 + 0.62 \, \Delta K_{US}$$
$$(4.35)^{**}$$
$$R^2 = 0.76 \qquad \text{Standard error/mean} = 0.11$$

where ΔX is the growth of exports in research intensive activities and ΔK_{US} is the growth of the US capital stake in manufacturing. The figures in parentheses are the t values of the respective regression coefficients, and the degrees of significance are denoted by * at the 5% and ** at the 1% levels.

This relationship has become stronger and statistically more significant in the later years. One reason is that foreign subsidiaries which initially begin by selling relatively new products may move towards a more export-oriented strategy in the course of their development; and, probably of greater importance, any such trend is complemented by the growth of intra-firm exports associated with the wider role played by rationalised investment. Finally, indigenous firms might be expected to expand their exports as part of a competitive response to the increasing presence of US foreign

Table 4.3   The sales of US manufacturing affiliates in Europe and exports from the USA to Europe, 1957–1975 ($m)

		Europe			EEC six			West Germany		
		1957	1966	1975	1957	1966	1975	1957	1966	1975
Chemicals	(a)	822	3,417	19,798	275	1,733	12,478	46	464	3,063
and	(b)	353	927	2,666	223	596	1,816	43	121	352
Allied	(c)	2.33	3.69	7.43	1.23	2.91	6.87	1.07	3.83	8.70
Primary and	(a)	435	1,619	7,985	145	648	4,458	45	207	2,057
Fabricated	(b)	462	497	1,220	214	296	633	70	87	189
Metals	(c)	0.94	3.26	6.55	0.68	2.19	7.04	0.64	2.38	10.88
Mechanical	(a)	1,009	4,099	21,235	502	2,431	13,007	228	903	4,571
Engineering	(b)	567	1,571	5,451	306	822	2,923	59	238	825
	(c)	1.78	2.61	3.90	1.64	2.96	4.45	3.86	3.79	5.54
Electrical	(a)	678	2,170	11,891	299	1,214	8,463	73	N.A.	3,133
Equipment	(b)	114	644	2,179	55	338	1,202	8	100	421
	(c)	5.95	3.37	5.46	5.44	3.59	7.04	9.13	N.A.	7.44
Transportation	(a)	1,700	5,012	16.339	764	2,747	9,822†	N.A.	N.A.	5,917
Equipment	(b)	244	647	2,493	130	350	1,152†	14	138	419
	(c)	6.97	7.75	6.55	5.88	7.85	8.53†	N.A.	N.A.	14.12
Rubber	(a)	262	662	2,190	77	276	1,321	N.A.	N.A.	204
Products	(b)	97	128	178	60	86	118	19	29	25
	(c)	2.70	5.17	12.30	1.28	3.21	11.19	N.A.	N.A.	8.16
Paper	(a)	34	384	3,221	13	241	1,671‡	6	61	420
and	(b)	91	257	916	44	157	482‡	15	55	195
Allied	(c)	0.37	1.49	3.52	0.30	1.54	3.74‡	0.40	1.11	2.15
All	(a)	4,940	17,363	82,650	2,074	9,290	52,221	890	3,920	19,365
Products	(b)	1,928	4,671	15,103	1,032	2,645	8,326	228	768	2,426
	(c)	2.56	3.72	5.47	2.01	3.51	6.27	3.90	5.10	7.98

(a) Sales of affiliates; (b) US exports; (c) Sales/export ratio.
N.A. Not available
*   Excluding paper and allied.
†   Excluding Netherlands transportation equipment.
‡   Excluding Netherlands paper and allied.
§   Excluding rubber products and transportation equipment.
Source: United Nations, *Commodity Trade Statistics*, 1957, 1966 and 1975; US Department of Commerce, *Survey of Current Business*, 1960 (supplement), August 1974 and February 1977.

Table 4.3 (continued)

France			Benelux			Italy			UK		
1957	1966	1975	1957	1966	1975	1957	1966	1975	1957	1966	1975
114	475	2,907	60	478	4,771	55	316	1,736	517	1,365	4,738
48	98	278	86	306	953	46	71	233	44	171	374
2.38	4.85	10.46	0.70	1.56	5.01	1.20	4.45	7.45	11.75	7.98	12.67
60	154	680	19	116	1,135	21	172	584	230	781	2,630
53	74	125	45	56	215	46	79	103	128	119	352
1.13	2.08	5.44	0.42	2.07	5.28	0.46	2.18	5.67	1.80	6.56	7.47
192	810	3,686	36	386	3,090	46	333	1,660	480	1,479	6,623
113	248	813	72	210	883	62	125	401	92	316	1,106
1.70	3.27	4.53	0.50	1.84	3.50	0.74	2.66	4.14	5.22	4.68	5.99
79	N.A.	2,194	67	N.A.	1,978	80	303	1,159	270	N.A.	1,737
13	113	279	16	68	285	18	58	217	9	129	428
6.08	N.A.	7.86	4.19	N.A.	8.94	4.44	5.22	5.34	30.00	N.A.	4.06
174	N.A.	2,528	138	N.A.	732†	N.A.	63	645	789	N.A.	4,070
39	63	245	67	130	307†	10	20	179	19	58	352
4.46	N.A.	10.32	21.06	N.A.	2.38†	N.A.	3.15	3.60	41.53	N.A.	14.12
N.A.	104	426	28	N.A.	596	N.A.	29	95	139	268	574
19	23	31	12	23	51	10	11	11	19	12	26
N.A.	4.52	13.74	2.33	N.A.	11.69	N.A.	2.64	8.64	7.32	22.33	22.08
N.A.	60	475	4	67	446‡	N.A.	54	330	21	89	449
12	31	112	11	31	61‡	6	39	115	41	65	193
N.A.	1.94	4.24	0.36	2.16	7.31‡	N.A.	1.38	2.87	0.51	1.37	2.33
619	2,497	12,895	353	1,603	13.752	202*§	1,269	6,209	2,446	6,511	21,721
266	650	1,883	309	824	3,118	172*§	403	1,259	352	870	2,831
2.33	3.84	6.85	1.14	1.95	4.41	1.17*§	3.15	4.93	6.95	7.48	7.67

affiliates, and this only really got under way later in the period. However, the issue remains as to the extent to which the general export growth was a direct result of the expansion of US affiliates or the expansion of other firms within the European countries.

The contribution of US affiliates to the changing share of European exports is shown in Table 4.4, which classifies European exports into those supplied by US affiliates and those supplied by other firms in Europe (including the affiliates of non-US foreign firms). It reveals that, in 1957, US affiliates accounted for 3.8% of all European plus US manufacturing exports; this compares with

63.8% for other firms in Europe and 32.3% for firms located in the USA (including the affiliates of non-US MNCs). Between 1957 and 1975, Europe's share of total European and US exports of manufactures rose from 67.7% to 77%. Of this increase of 9.3%, over half (5%) was due to an increase in the exports of US-owned foreign affiliates in Europe. An even more striking way of stating this is that while the share of European exports of non-US firms by 1975 had risen 6.9% above their 1957 share (4.4%/63.8%), the equivalent share of US foreign affiliate exports increased by 131.6% (5%/3.8%).

Treating the exports of US affiliates as a part of those of their parent group, the share of US firms in total US and European exports slipped back by just 4.4%, from 36.2% in 1957 to 31.8% in 1975, whereas exports from the USA fell by 9.3%, from 32.3% to 23% over the same period. In the food and paper and allied products sectors the share of US firms, as opposed to the share of the USA, did not fall at all. In other words, a large part of the catching up of European exports was accounted for by US firms switching the location of their production to Europe, as postulated in the model of Chapter 3.

For individual industries, the role of US affiliates varies considerably. In food, chemicals, mechanical engineering and paper products, the exports of affiliates rose much faster than those of other European-based firms between 1957 and 1975. In rubber products, exports from both affiliates and indigenous firms have increased as against exports from the USA which have remained stagnant, and there has also been something of a general shift towards European locations in the metals and electrical equipment industries. In the case of transportation equipment the share of US affiliates has fallen as this is the one sector in which the share of non-US firms has really taken off.

With one or two exceptions, the relative contribution of US foreign affiliates to the growth of European exports has been fairly consistent over the period, and indeed (apparently contrary to what was supposed in the model) US firms enjoyed a slightly higher share of the additional exports from Europe in the second ten years than the first (12.3% in 1965–75 compared with 10.7% in 1957–64). One reason is that they began from a higher base share in 1965 than in 1957. Another possible explanation is the trend towards more rationalisation of the European activities of

Table 4.4 The contribution of US affiliates in Europe to total European and US exports of selected manufactured products, 1957–1975 (%)

	1957				1965				1975			
	(1) Other firms in Europe	(2) US affiliates in Europe	(3) US	(4) (3)+(2)	(5) Other firms in Europe	(6) US affiliates in Europe	(7) US	(8) (7)+(6)	(9) Other firms in Europe	(10) US affiliates in Europe	(11) US	(12) (11)+(10)
Food Products	63.4	0.6	36.0	36.6	63.5	1.0	35.5	36.5	64.2	2.0	33.8	35.8
Chemicals and Allied	64.3	3.1	32.6	35.7	67.6	7.0	25.4	32.4	67.2	14.2	18.6	32.8
Primary and Fabricated Metals*	81.8	1.8	16.4	18.2	84.0	2.7	13.3	16.0	85.9	3.5	10.6	14.1
Mechanical Engineering	56.6	2.7	40.7	43.4	63.0	6.1	30.9	37.0	62.1	11.4	26.5	37.9
Electrical Equipment	63.7	3.4	32.9	36.3	69.7	5.1	25.2	30.3	71.7	6.5	22.4	28.9
Transportation Equipment‡	46.0	16.3	37.7	54.0	58.7	16.8	24.5	41.3	62.9	12.1	25.0	37.1
Rubber Products	58.8	7.4	33.9	41.2	65.1	14.4	20.5	34.9	70.8	16.6	12.6	29 2
Paper and Allied	78.9	0.5	20.6	21.1	75.8	0.8	23.4	24.2	77.8	6.3	15.9	22.2
All Products	63.8	3.8	32.3	36.2	67.3	6.1	26.6	32.7	68.2	8.8	23.0	31.8

* Excluding non-ferrous metals.
‡ Only motor vehicles.
Source: as for table 4.3.

US firms, with its consequent increase in intra-group trade. Also, as pointed out in Chapter 3, the general tendency towards an increased internationalisation of production has meant that the firms of all countries have expanded their international production relatively faster compared to the growth of their domestic exports, and consequently foreign affiliate export shares may give a misleading impression of the overall competitive performance of the MNCs concerned, especially later in the period.

This helps to explain why, in Table 4.4, the share of European exports accounted for by US affiliates increased by as much as it did. The revival of indigenous European firms and their reaction to the presence of US affiliates by increasing their rate of technological accumulation cannot be fully appreciated from this evidence. However, as suggested in Chapter 3, it is clear that an indigenous revival occurred in some sectors, such as transportation equipment, but not in others. More illuminating in this respect is Table 4.5, which sets out details about the individual countries in Europe. Here the striking feature is the distinction between large and small industrialised economies. While for the UK, West Germany, France and Italy, the increase in exports by indigenous firms between 1966 and 1975 paralleled that of US affiliates, for Belgium and the Netherlands it was considerably less pronounced.

The competitive response to the 'American Challenge' was strongest amongst the firms of the largest industrialised countries, the UK, West Germany, France and Italy in the second decade of the period. While the share of US affiliates in total European exports rose sharply in all countries from 1957 to 1966 (from 3.8% to 7.0% in West Germany, 1.0% to 6.3% in France, 2.3% to 10.6% in Belgium and Luxembourg, and 11.7% to 17.9% in the UK), this was not repeated in the larger European economies over the following nine years. As can be seen from Table 4.5, in 1966–75 the US affiliate share of European exports rose from just 7% to 7.4% in West Germany, 6.3 to 8% in France, 4.4% to 4.9% in Italy, and 17.9% to 19.7% in the UK, despite the sharper increases from 7.8% to 13.1% in the Netherlands and from 10.6% to 18.9% in Belgium and Luxembourg. Indigenous firms performed especially well in transportation equipment and food products in West Germany, in non-electrical machinery in France, in non-electrical and electrical machinery and food products in Italy, and in rubber tyres, electrical machinery, transportation

Table 4.5   The share of majority-owned affiliates of US firms in the total exports of European countries, 1957–1975 (%)

	Europe			EEC six		West Germany		France		Italy		Netherlands		Belgium and Luxemburg		UK	
	1957	1965	1975	1966	1975	1966	1975	1966	1975	1966	1975	1966	1975	1966	1975	1966	1975
Food Products	0.9	1.6	3.1	3.0	3.5	6.4	4.0	1.2	1.4	2.5	1.2	4.4	5.6	2.6	5.3	9.4	2.3
Chemicals and Allied	4.6	9.4	17.5	7.9	18.3	2.0	5.3	7.1	13.3	5.8	11.7	17.8	39.6	28.4	38.7	16.2	26.1
Primary and Fabricated Metals	2.1	3.1	3.9	1.2	2.2	1.1	1.4	0.6	3.4	1.8	2.8	5.9	4.3	0.4	2.5	9.1	10.8
Mechanical Engineering	4.5	8.9	15.5	11.6	14.0	6.7	7.9	22.9	14.5	8.4	7.7	N.A.	22.3	37.3	72.1	20.6	31.9
Electrical Equipment	5.1	6.8	9.0	7.2	9.8	4.4	7.6	7.6	7.8	13.6	10.0	2.7	3.0	14.5	39.1	20.3	12.7
Transportation Equipment	26.1	22.2	16.1	15.8	13.2*	23.7	20.2	1.9	6.9	N.A.	1.7	N.A.	N.A.	20.4	N.A.	32.8	29.9
Rubber Products	11.1	18.1	19.0	18.1	22.9	0.0	3.0	17.9	27.2	N.A.	3.7	N.A.	N.A.	57.7	N.A.	21.0	20.0
Paper and Allied	0.6	1.1	7.5	6.2	6.6†	1.1	2.6	3.4	4.4	18.3	24.4	12.1	N.A.	8.9	N.A.	2.6	7.2
Other Manufacturing	N.A.	3.8	3.8	3.0	3.3	2.8	4.2	3.3	6.9	1.6	1.3	5.3	1.7	3.1	2.0	9.3	10.7
All products	5.6‡	7.2	9.8	7.0	9.4	7.0	7.4	6.3	8.0	4.4	4.3	7.8	13.1	10.6	18.9	17.9	19.7

* Excluding Benelux transportation equipment.
† Excluding Benelux paper and allied products
‡ Excluding 'other manufacturing'.
N.A. Not available

Source: US Department of Commerce, Survey of Current Business, 1960 (supplement), October 1970, and February 1977; United Nations, Commodity Trade Statistics, 1965 and 1975; US Department of Commerce, Tariff Commission Report, 1973.

equipment and food products in the UK. This tends to confirm the impression of Table 3.1 in the case of Italy and West Germany (except for food products), but it suggests a different sectoral pattern for the indigenous firm revival in France and the UK.

Taken together, Tables 3.1 and 4.5 show that while in the case of the larger industrialised countries the presence of US affiliates tended to act as a spur to the technological capacity of indigenous firms, in the smaller countries indigenous competition has been inhibited. This is closely connected with the breadth of the underlying research strengths of European companies, and their capacity for technological accumulation across a range of activity in the rather widely defined industrial categories. The larger countries are, as a rule, home to a greater number of the strongest firms with broad technological capabilities in their respective industries. It is worth noting that even in the smaller European countries a clear response on the part of local companies can be seen in sectors in which they have been technologically specialised (see Table 2.3). In the case of the Netherlands the revival came in the electrical equipment sector, while in Belgium and Luxembourg this effect was witnessed in other manufacturing (which in Table 4.5 includes non-metallic mineral products, textiles, and professional and scientific instruments).

This suggests the conclusion that it is right to suppose that it is the technological capacity of indigenous firms that was the major factor in determining the success of the European corporate response to the expansion of innovative firms, even if the size of local markets made an additional contribution. In the biggest European countries local markets were sufficiently large to allow both foreign and indigenous firms to produce at an economic scale, particularly in industries in which there were substantial scale economies. However, given trade liberalisation within the EEC in the period in question, this is likely to have been only a secondary influence.

Though the product classification adopted is too broad for any firm conclusions about the technological impact of US firms in Europe, the view that revival depended upon traditional technological strengths is supported by studies carried out at a rather more disaggregated level.

Illustrative of this is research conducted by Burstall, Dunning and Lake (1981) on the impact of MNCs on the technological capacity of the pharmaceutical industry in OECD countries. The study quite clearly demonstrated that whereas in the UK with its thriving indigenous pharmaceutical sector, foreign (and particularly US) firms have had a stimulating effect on indigenous technological innovation – and hence the international competitive position of UK-owned pharmaceutical companies – the smaller European countries which lacked any such tradition, especially the Scandinavian countries and Ireland, relied almost completely on foreign firms for the import of both innovatory capacity and quite often of production technology associated with the intermediate and final processes of drug manufacture.

A related report on the impact of MNCs in the food processing industry (OECD, 1979), also confirms that following the rapid expansion of US food companies in Europe in the 1960s, that from around 1968 certain European companies, and in particular large UK firms, began to respond through their own multinational growth, based largely on the development of new products and the diversification into new food sub-sectors. The strongest UK food companies are highly competitive, and they have historically had extensive networks of international production (Cantwell and Dunning, 1985). Lake's (1976b) study of the semiconductor industry also suggested that amidst rapid technological change spreading out from the USA, experienced European firms and particularly the established corporations of the largest industrialised countries had been best placed to retain their position, and initiate new innovations of their own.

Similar evidence on the competitive impact of the newly emerging Japanese investment within the UK has recently been collected by Dunning (1986). He describes how in the consumer electronics sector, Thorn-EMI has improved its product designs, work practices, and quality control procedures as a result of the entry of Japanese-owned affiliates producing colour televisions. However, this is a sector in which British companies have a lot of ground to catch up on the Japanese, and it may be of greater interest to see how the Dutch firm Philips responds. A clearer case of a virtuous circle in the case of the UK has been the pharmaceuticals industry, which is directly related to the role of US inward investment (Lake, 1976a). The R&D of local UK firms has increased tremendously since the major wave of US investment in the 1960s, as well as the location of R&D facilities by the foreign affiliates of MNCs (see

Brech and Sharp, 1984), and the technological capacity of the UK sector has benefited accordingly.

By contrast, the UK motor vehicle industry is an example of a sector in the midst of a vicious circle of cumulative decline, as represented by the continuing loss of markets by local firms since the 1960s, and an increasing dependence upon external sources of supply. This is in part due to the rising global integration of the industry, in which UK affiliates have been linked into MNC networks whose major technological capacity lies outside the UK (see Foreman-Peck, 1986). There are genuine fears in this case that Japanese investment in the UK will act as a 'Trojan horse', with affiliates concentrating on assembly-type activities, and importing research-intensive components from abroad.[6]

4.4   The Role of Licensing in the European Revival

According to the model of Chapter 3, where US firms were technologically strong they might have been expected to expand their interests in Europe through the licensing of non-affiliated firms, as well as through the growth of affiliates. However, production within US MNCs expanded much faster than production of European firms licensed by independent US firms. This is illustrated by comparing the growth of the sales of US MNCs with that of licensing fees from unaffiliated foreigners, on the assumption that licence fees represent a roughly constant proportion of sales in the period concerned. The total sales of US affiliates in Europe grew by 367.3% between 1966 and 1975 (and by 458.2% in the EEC six and 217.3% in the UK); while US receipts of royalties and fees from unaffiliated European residents rose by 126.2% between 1967 and 1977 (and by 111.2% in the EEC six and 49.1% in the UK).[7]

Table 4.6 sets out cross-industry evidence on the growth of receipts and payments of royalties and fees in manufacturing industry between unaffiliated firms for three European countries and the USA, between 1965 and 1975. The fastest rates of growth of both receipts and payments have occurred in the European countries (particularly in West Germany), and for no country was there a really substantial divergence between the growth of total receipts and payments. These two developments in non-affiliate

Table 4.6 The growth of receipts and payments of royalties and fees from non-affiliate licensing, 1965–1975 (1965=100) of the USA and selected European countries

	USA		UK		Sweden		West Germany	
	Rec.	Pay.	Rec.	Pay.	Fec.	Pay.	Rec.	Pay.
Food Products	137.5	133.3	296.5	485.7	200.0	550.0	266.6	480.0
Chemicals and Allied	174.3	151.9	235.1	488.8	381.8	321.4	426.4	285.0
Primary and Fabricated Metals	162.1	33.3	2300.0	372.7	116.6	241.2	1366.7	341.7
Mechanical Engineering	223.3	275.0	328.9	141.1	407.1	413.3	451.2	453.1
Electrical Equipment	215.6	500.0	375.3	495.7	314.3	176.9	398.6	426.2
Transportation Equipment	190.0	266.7	73.1	31.6	14.3	925.0	241.8	168.3
Textiles and Clothing	21.4	100.0	32.4	237.5	150.0	700.0	250.0	200.0
Paper and Allied	350.0	200.0	863.2	643.5	350.0	183.3	433.3	900.0
Other Manufacturing	438.5	187.5	1108.5	478.7	357.1	500.0	896.0	766.6
Total Manufacturing	203.0	177.1	322.0	327.5	297.9	346.3	402.0	347.0

Source: Clegg (1987). Where possible the data conform to the US definition of royalties and fees.

licensing in the 1965–75 period are both further indications of a technological revival on the part of indigenous European firms after the mid-1960s, especially amongst the larger firms of the major industrialised countries. The most pronounced improvement in Europe's net technological viability was recorded by the chemicals industry; indeed in this sector, by 1975, the royalty receipts of West German-based firms, from unaffiliated foreign residents, were equal to those of US-based firms.

This is interesting, as the chemicals sector is one in which the share of US affiliates in total European exports increased between 1965 and 1975 (see Table 4.5). This may partly reflect the fact that European, and especially West German, firms increasingly came to rely on international production and licensing agreements with foreign firms as a means of exploiting their technological advantages abroad, but it also seems to reflect an increase in intra-industry production between Europe and the USA. In terms of the taxonomy of Chapter 3, the chemicals industry appears to be representative of an industry type (iii), in which both US and European firms have advantages.

In the sector in which the export performance of West German-based firms improved *vis-à-vis* the affiliates of US firms, that is transportation equipment, the royalty receipts of West German firms clearly rose faster than their payments. This lends further support to the view that there was an increase in the rate of technological accumulation on the part of such indigenous firms. In West Germany at least, motor vehicles have been a type (ii) sector, in which local firms have proved to be strong in innovation by international standards.

However, in the case of food products in West Germany, royalty payments have risen markedly faster than receipts, though both remain at a comparatively low level. This suggests that at least part of the higher export share of West German firms in food products has been due to the efforts of companies unaffiliated to but licensed by foreign firms. A similar pattern emerges in the case of the UK, where it appears that the local gains in the electrical equipment industry owe much to non-affiliate licensing. The same qualification about the relatively fast export growth of local firms is less likely to apply in the case of food products, in which sector British companies have long been heavily dependent upon international production and therefore have less reason to license unaffiliated firms abroad.

4.5   Conclusions

This chapter has served to put the flesh on the bones of the model of the last chapter. The transition from stages 3 and 4 of that model to stages 5 and 6 was made in the mid-1960s. Before that time European catching up had been based in large part on the growth of the international production of US firms in Europe. After that time, although there were some sectors (type (i)) in which the exports and production of foreign affiliates continued to increase their share and to rise to a peak, there were other sectors in which the rise in the share of US foreign affiliates was halted (type (iii)), and still others in which their share actually slipped back (type (ii)). The division between these three different types of industry varied between European countries.

There were more sectors of types (ii) or (iii) in the larger European countries whose firms had a broader degree of technological specialisation, and were therefore proved capable of a wider and more comprehensive response to the 'American Challenge'. However, even in the smaller European countries there were some areas of traditional technological strength in which the gains of US foreign affiliates proved to be rather short-lived. In such cases the entry and expansion of US-owned affiliates in European production and exports was highly beneficial, in that they provided a competitive spur to indigenous firms. The competitive stimulus of the direct entry of US firms into Europe revitalised underlying technological skills where they existed.

The factor which has been only briefly mentioned in the empirical work of this chapter is the same as that missing from the model of Chapter 3 on which it has been based; namely, that given a general trend towards the internationalisation of industry, the extent or otherwise of a European revival cannot be fully assessed by looking simply at the relative shares of US affiliates and indigenous firms of markets served from a European production base. By the early or mid-1970s European firms had begun to increase their international production quite rapidly. To take the argument further it is necessary to consider the world market shares of US and European firms served from any production location in the more international technological competition that has prevailed since the early 1970s. The competitive position of US, European and Japanese firms in total international economic activity in more recent years, and how this relates to the pattern of their technological accumulation, is discussed from Chapter 5 onwards.

NOTES

1 The empirical information used in the tables accompanying this chapter was first compiled for a co-authored paper, Dunning and Cantwell (1982).
2 The more rapid growth of exports from EEC countries compared with those in the rest of Europe is at least partly the result of the increase in intra-EEC trade following the reduction in tariff barriers since 1958.
3 For another account see also Dunning (1972).
4 A review of some such surveys can be found in Bergsten, Horst and Moran (1978). See also Hood and Young (1980).
5 West Germany, France, Italy, Netherlands, Belgium and Luxembourg, the UK, Denmark and other Europe.
6 For a further discussion of the contrast between the pharmaceuticals and motor vehicle industries in the UK in the context of the relationship between the local capacity for technological accumulation and the existence of a mechanism of cumulative causation in responding to foreign competition see Cantwell (1987).
7 The data on which these calculations are based can be found in Kroner (1980).

5

The European and Japanese
Response to the International
Expansion of US Manufacturing
Firms

5.1   Introduction

Chapters 3 and 4 have described how, underlying the 'catching up' of European export shares in international markets for manufactured goods between the mid-1950s and the mid-1970s, innovative US firms expanded their operations in Europe and helped to spur an indigenous revival in areas of traditional technological strength. However, from the early 1970s onwards this revival increasingly took the form of an expansion of the international production of European firms rather than of exports from their home countries. The trend towards the internationalisation of production became generalised across the firms of all countries. Firms that had previously served international markets principally from domestic production sites increasingly attempted to 'catch up' with the global networks of their more mature MNC rivals.

Moreover, by now it had become evident that it was not only Europe that was 'catching up' the USA in exports of manufactures, but Japan as well. By the late 1970s this also fed through into a rapid internationalisation on the part of many Japanese companies. Just as innovative US firms moved from domestic exports to international production in the other major industrialised countries in the 1950s and 1960s, so Japanese, West German, Italian and

French companies have done likewise in the 1970s and 1980s. As their rate of technological accumulation caught up with and in some cases surpassed that of their US competitors, they captured international market shares, and found that such growth was most easily sustained by establishing an international network of productive activity (Cantwell, 1989). One reason for this is that carrying out research and production in a variety of international centres of innovative activity increases the capacity for and the complexity of technological accumulation within MNCs. A further reason is that faced with the rapid export growth of a group of innovative companies, weaker firms abroad try to persuade governments to erect trade barriers in response, and this may have played a role in the promotion of international production by the newer Japanese MNCs.

The outward push of the most innovative firms has moved technological competition to a more genuinely international arena, in which the chief actors are today members of global oligopolies. The internationalisation of production has not been confined to the leading MNCs, however. It has extended even as far as Third World firms that have simpler and more limited forms of technological accumulation (Lall et al., 1983; Cantwell and Tolentino, 1987). One factor in the general trend towards internationalisation has been a continuing fall in transport and communications costs, which has increased the scope for international specialisation by encouraging intra-firm trade and the movement of personnel between countries. This has been associated with organisational innovation that has lowered the costs of managing more complex multi-plant operations within the firm (Casson, 1983), and which has facilitated the strategic coordination of technology creation across national boundaries.

In the past, the general trend towards internationalisation has been at its strongest in periods of high innovation in the world economy or the industrialised countries as a whole (Schumpeterian upswings) such as 1890–1914 or 1950–70. This is related to the ability of innovative leaders, such as US firms after the war, to capture international market shares and to establish multinational networks as a means of consolidating rapid growth. However, the general trend towards internationalisation has continued unabated, despite the fall in world patenting activity and growth rates since the early 1970s. Technological competition has been sustained by means of the internationalisation of research and production, and the increasing technological diversity of firms that has accompanied it.

The creation of an international division of labour within the firm has become a prerequisite for the survival of modern MNCs, enabling them to specialise in particular activities in the most suitable locations, and to coordinate a decentralised research strategy. This has widened their capacity to innovate, and to diffuse new technological advances more rapidly to all parts of their network, enhancing further technology creation as well as production. It has also increased their ability to enter into co-alitions with other innovative firms to their mutual benefit, as the pattern of their technological accumulation overlaps and becomes more complementary.

A further consideration is the characteristics of the newly emerging technology paradigm (Freeman and Perez, 1989) in which electronics plays a leading role. Although the general industrial application of the new technologies is still at an early stage, they have had much more immediate effects in the communications and telecommunications sector. The costs of the transfer of information and international communications within the firm have continued to fall. The new technologies and methods of work are more suited to a greater international dispersion of productive activity for another reason as well. Whereas the major technologies of the 1950s and 1960s were essentially scale-intensive in their applications in the chemicals, motor vehicles, metals, aircraft and other manufacturing industries, the newer technologies need not be. By reorganising themselves around the development of core skills, firms today have become much more dependent upon economies of scope, and less dependent upon economies of scale. This is particularly true of larger MNCs, and it has undermined the major motive for geographical concentration.

An additional explanation of why firms may prefer to move to a wider dispersion of production units is that it weakens the bargaining power of trade unions based on the organisation of closed shops in large plants. The trend towards internationalisation may therefore increase the share of profits in the total revenue of firms (Cowling, 1986). There are also political reasons why the pace of internationalisation has not slowed down in the 1970s and 1980s in the way that it did in the inter-war period. The revival of nationalistic movements and protectionist pressures has been much more limited and constrained by the fears of an escalating trade war. As a result, the political risk associated with international production has not changed very much in the industrialised countries, unlike in the 1930s.

As a consequence of the steady internationalisation of production by non-US firms since at least the early 1970s, success and failure in technological competition at a corporate level has to be measured by shares of total international economic activity, not only export shares. To this end, some new and original data on the international production of manufacturing firms of the industrialised countries (and of the world as a whole) were derived, and are presented in Section 5.2. By estimating the value of exports for which foreign-owned affiliates producing in each host country are responsible (a component which is included in total international production), it is then possible to calculate the value of the domestic exports of indigenous firms. Such international production and indigenous firm exports combined represent the total international involvement of each country's firms, except for non-affiliate licensing and other contractual arrangements. Unfortunately, non-affiliate licensing had to be excluded from this part of the study, due to the lack of industrially disaggregated data, especially in the case of France and Italy. However, the analysis retains much of its validity despite this omission, since although production under foreign licence is important in some sectors, it is still on the whole small by comparison with international trade and production (though see Clegg, 1987, for a fuller discussion of the role played by non-affiliate licensing).

The focus is on the six major industrialised countries – the USA, West Germany, the UK, Italy, France and Japan – and their firms. These are the groups that have been most heavily involved in technological competition at an international level. The intention of this chapter is to describe and present the data on which the study of subsequent chapters is based, with a preliminary exploration of what they show. Having looked at the data for the year 1982 in Section 5.2, Section 5.3 proceeds to examine changes in the international market shares of companies over the period 1974–82. This reveals the extent and the industrial composition of the European and Japanese response to the earlier international expansion of US firms. The point particularly at issue is the degree to which this depended upon the established position of European and Japanese firms, and upon their pattern of technological accumulation. Section 5.4 extends the argument through a discussion of the relationship between the sectoral pattern of the success and failure of national groups of firms and the performance of their countries in international trade.

5.2 The Significance of International Production and Exporting as Means of Serving International Markets

While the six major industrialised countries accounted for 57% of world exports of manufactured goods in 1982, the estimates displayed in Table 5.1 suggest that their firms were responsible for 83% of the sales from international production in manufacturing in the same year. As in Chapter 4, international production is measured by the value of sales that derives from it, which makes it comparable with data on export sales. The equivalent figures for 1974 show that the six countries held 55% of world exports and their firms 81.5% of international production. This immediately demonstrates that the competitive position of the firms of these countries is somewhat stronger than measures of the competitive position of the countries themselves would indicate.

The estimates of international production represent for the most part original calculations, and they have been derived from a variety of national sources, which provide data on the value of sales by the foreign affiliates of their MNCs, and on their outward foreign direct capital stock (see Dunning and Cantwell, 1987). The former can be used directly, given that as just mentioned international production has been measured by sales, which approximate to gross production or output (but not net output or value added). Where the latter (foreign direct capital stock data) has been used it has been necessary to estimate the industrial variation in the ratio of the stock of FDI to international production. This has been achieved by applying an average industrial variation derived from the equivalent ratios of those countries for which data on both foreign capital stock and international production were available. In addition, since foreign capital stock is calculated at historic cost, its value relative to foreign production varies substantially between countries. In general, due to the rise in

Table 5.1   The geographical distribution of international production
in manufacturing by source country, 1982

Source country	Value of production ($m)	Share of production (%)
USA	357,244	42.43
West Germany	89,586	10.64
UK	154,230	18.32
Italy	13,141	1.56
France	45,333	5.38
Japan	41,047	4.87
World total	842,027	100.00

Source: Various national sources as identified in Dunning and Cantwell (1987), and aggregate firm level data used in Dunning and Pearce (1985). In the case of West Germany, international production data (or more accurately on the sales of West German-owned foreign affiliates) is available directly from Deutsche Bundesbank, 'Die Kapitalverflechtung der Unternehmen mit dem Ausland nach Ländern und Wirtschaftszweigen 1978 bis 1983', Supplement to *Statistische Beihefte zu den Monatsberichten der Deutschen Bundesbank*, series 3, Zahlungsbilanzstatistik, No. 3, March 1985. Similar data are available for the USA for 1977 in US Department of Commerce, US Direct Investment Abroad, 1977. An equivalent set of estimates for 1982 were made by applying the proportional change in the investment position between 1977 and 1982, using data on the outward direct investment position of the USA in 1982 taken from US Department of Commerce, *Survey of Current Business*, August 1984. For other countries the primary evidence from official national sources is on the stock of foreign direct investment, whether in the form of a survey of foreign capital stock, or the cumulative investment flow position. For the UK the principal source used was Department of Trade and Industry, Business Monitor M4 Supplement, *Census of Overseas Assets, 1981*, which was used to update the estimates obtained in Cantwell (1984). For Italy the official source on direct investment stock was Bank of Italy, *Annual Report, 1984*. This was supplemented by survey evidence on direct investment and international production made available to the author by research teams at the University of Rome (see Acocella, 1985), and at the Institute for Social Research in Milan (see F. Onida, ed., *L'Internazionalizzazione del Sistema Industriale Italiano*, Bologna: Il Mulino, 1986). In the case of France, cumulative direct investment flows were calculated from a series of issues of Bank of France, *Annual Report, Annexe 2*, 'Mouvements de capitaux á long terme entre la France et l'extérieur'. This was combined with survey evidence on the international production of 182 French multinationals as cited in Savary (1984). The main source of information on Japan's cumulative direct investments flows, and on Japanese-owned foreign affiliate surveys, is the Ministry of International Trade and Industry; see MITI, *Overseas Business Activities of Our National Enterprises*, 1983, and *News from MITI*, 'Japan's direct overseas investment in FY 1983', 16 July 1984. For a discussion of the methods used see the text and Cantwell (1984).

inflation since the late 1960s, where a country's foreign capital stock is of a longer vintage (as in the case of the USA and the UK), its true value will be underestimated by a greater amount. The different ratios between FDI and foreign production across countries have been calculated using a technique for the revaluation of foreign capital stock, explored further using the example of the UK by Cantwell (1984). A further discussion of the issues involved in revaluing foreign capital stock data can be found in Cantwell (1986b).

These estimates have been supplemented by information on the international production of the world's 792 largest industrial firms, which has been grouped collectively by the country of origin and the industry of each firm by Dunning and Pearce (1985). Where, occasionally, the estimates of international production for country – industry combinations described above fell below the equivalent value obtained from the sample of the largest firms, then the estimate was raised. The value of international production for the world as a whole, and its industrial distribution, was calculated essentially from the Dunning and Pearce sample. They claim (and it is here assumed) that the largest 792 industrial firms were responsible for about 85% of total international production in manufacturing in the world in 1982. The industrial distribution of total international production was then adjusted to take account of discrepancies in the sectoral pattern of the total international production of the six countries being considered, when compared with the sectoral pattern of the international production of their leading firms calculated by Dunning and Pearce. Such discrepancies arise in part due to slight differences in the sectoral classification adopted by Dunning and Pearce and that in wider use in official statistics, but they are also due to the different sectoral distribution of the international production of smaller MNCs.

Data on the exports of foreign affiliates by industry from within each host economy were obtained by a similar but lengthier route. Where they were not available directly, figures on the production of foreign-owned affiliates within each country were obtained by using the same method as applied in calculating the international production of the country's own firms (from an estimated proportional relationship with their stock of FDI). The total world output of foreign affiliates across all host countries is, of course, equal to the total international production of the firms of all home

Table 5.2   The geographical distribution of international production and indigenous firm exports in manufacturing combined, by source country, 1982

Source Country	Value of production plus exports ($m)	Share of production plus exports (%)
USA	521,269	27.60
West Germany	230,466	12.20
UK	212,816	11.23
Italy	78,434	4.15
France	121,709	6.45
Japan	176,687	9.36
World total	1,888,357	100.00

Source: As for table 5.1, and United Nations, *Yearbook of International Trade Statistics*, 1984. The US exports of foreign-owned affiliates are directly available from US Department of Commerce, *Survey of Current Business*, December 1984. Data on the total sales of foreign-owned affiliates are available for West Germany (in the source cited under Table 5.1), and the UK, in Department of Trade and Industry, *Census of Production, Summary Tables*. For France and Italy data on inward direct investment stocks were used, from the sources quoted above. Information on the sales and exports of foreign-owned affiliates in Japan in 1982 is available in *News from MITI*, 'Survey on foreign affiliates in FY 1983 (Summary)', 8 October 1984.

countries. Then, using data on foreign affiliate exports where they are directly available, an average propensity to export out of international production was estimated for each industry. This was then applied to production values where foreign affiliate exports were not directly available.

The value of exports of foreign affiliates was then deducted from the total value of exports, in order to assess the value of exports for which indigenous firms are responsible. When such domestically owned firms' exports are added to the international production of the same firms, the results for the year 1982 are those shown in Table 5.2. A similar procedure was applied to obtain estimates of international production and indigenous firm exports in 1974, which are used in later sections.

For the world as a whole, international production in manufacturing was about 45% of total international production plus domestic firms' exports in 1982. However, this understates the significance of international production as a means of serving world markets. This is because a certain proportion of both indigenous firm exports, and foreign affiliate exports (part of international production) represent the intra-firm trade of MNCs. Taking this into account, it seems reasonable to conclude that exports and international production are of more or less equal importance as means of servicing world markets for manufactured goods.

However, it is well known that the relative importance of international trade and production varies between firms of different nationalities. This is confirmed by a comparison between Tables 5.1 and 5.2. Firms from the USA and the UK are substantially more reliant on international production than the average. They account for 42.4% and 18.3% respectively of international production in world manufacturing (over 60% combined), but only 27.6% and 11.2% respectively of international production plus exports (less than 40% combined). West German, Italian, French and Japanese firms, though, are comparatively more dependent on exports, even if the evidence on the changing position since 1974 shows that all four groups are moving closer to the average as their firms undergo multinationalisation.

A similar analysis can be conducted with reference to the strategy of industrial groups of firms. International production is relatively more important than exporting compared with the average in the cases of food products, chemicals, electrical equipment, motor vehicles, rubber products, non-metallic mineral products, and coal and petroleum products (see Tables 5.3 and 5.4). The export route is relatively favoured in the cases of metals, mechanical engineering, other transport equipment (most notably aircraft), textiles, and other manufacturing (the coverage of which includes paper and wood products, and professional and scientific instruments). It must, of course, be remembered that these calculations are for the world as a whole (incorporating countries whose firms have little international production of their own), and that the industrial variation may differ for the firms of different countries. In particular, it depends upon the location advantages of the country in question, which, in sectors where they are strong, tend to encourage a higher proportion of local exports in the total activity of the country's firms.

Table 5.3   The industrial distribution of international production in manufacturing, 1982

Industry	Value of production ($m)	Share of production (%)
1. Food Products	132,373	15.72
2. Chemicals	119.281	14.17
3. Metals	57,056	6.78
4. Mechanical Engineering	59,532	7.07
5. Electrical Equipment	97,299	11.56
6. Motor Vehicles	85,665	10.17
7. Other Transport Equipment	9,039	1.07
8. Textiles	14,625	1.74
9. Rubber Products	20,958	2.49
10. Non-metallic Mineral Products	22,958	2.73
11. Coal and Petroleum Products	187,889	22.31
12. Other Manufacturing	35,352	4.20
Total world manufacturing	842,027	100.00

Source: As for table 5.1.

5.3 Changes in International Market Shares since 1974: the European and Japanese Corporate Response to American Multinationals

The firms of the six major industrialised countries were responsible for about the same share of total international production in manufacturing in 1974 and 1982 (81.5% and 83.2% respectively). However, as is clear from a comparison of Tables 5.5 and 5.6, this share of a little over 80% was very differently distributed in the 1980s than it had been ten years earlier. Just as European firms had begun to catch up with US MNCs in terms of export growth from Europe between the mid-1960s and mid-1970s, so after the mid-1970s they began to catch up in terms of their networks of international production. US firms slipped back from 55.3% of total international production in manufacturing in 1974 to only 42.4% in 1982. Over the same period, West German firms increased their share from 6.8 to 10.6%, British firms from 12.4 to 18.3%, Italian firms from 1.1 to 1.6%, French firms from 3.9 to 5.4%, and Japanese firms from 2.1 to 4.9%.

Table 5.4   The industrial distribution of international production and
indigenous firm exports in manufacturing combined, 1982

Industry	Value of production plus exports ($m)	Share of production plus exports (%)
1. Food Products	281,149	14.89
2. Chemicals	219,271	11.61
3. Metals	176,148	9.33
4. Mechanical Engineering	200,197	10.60
5. Electrical Equipment	202,591	10.73
6. Motor Vehicles	184,296	9.76
7. Other Transport Equipment	55,820	2.96
8. Textiles	107,817	5.71
9. Rubber Products	38,244	2.03
10. Non-metallic Mineral Products	38,839	2.06
11. Coal and Petroleum Products	225,971	11.97
12. Other Manufacturing	158,025	8.37
Total world manufacturing	1,888,357	100.00

Source: As for table 5.2.

US MNCs suffered their heaviest losses in their world position
in international production in the chemicals sector, in which West
German and British firms were well placed to respond; in non-
metallic mineral products, especially to French companies; in
electrical equipment to the Japanese; in motor vehicles to both
Japanese and European firms; and in coal and petroleum products
to British firms. In each of these cases gains in international
production were achieved on the basis of a relative strength in
technological accumulation from at least the 1960s onwards (see
Table 2.3). The only qualification needed here is that in the motor
vehicle industry British firms lost technological advantage and
market shares, while the leading Italian firm (Fiat) relied upon its
links with the traditionally strong Italian machine tool sector (for
a more detailed discussion of trends in this industry see Cantwell,
1987). Taking manufacturing industry as a whole, those MNCs
that had the fastest rates of technological accumulation overall,
the West Germans and Japanese, experienced the greatest proportional rise in their share of international production, taking

Table 5.5   National firms' shares of international production by manufacturing sector, 1974 (%)

	USA	West Germany	UK	Italy	France	Japan
1. Food Products	32.8	1.1	21.8	0.5	1.2	1.2
2. Chemicals	52.8	22.1	9.2	3.3	7.8	3.8
3. Metals	40.4	8.7	23.2	0.2	5.4	1.0
4. Mechanical Engineering	51.1	8.6	10.9	0.3	2.5	1.2
5. Electrical Equipment	65.6	13.2	10.8	1.0	3.8	4.6
6. Motor Vehicles	66.7	11.7	9.3	2.5	6.3	2.6
7. Other Transport Equipment	14.4	23.8	36.5	0.3	3.3	0.3
8. Textiles	40.5	10.3	40.1	0.4	2.4	5.3
9. Rubber Products	55.9	5.6	9.8	12.3	11.3	4.2
10. Non-metallic Mineral Products	47.6	5.3	19.9	0.1	11.1	4.2
11. Coal and Petroleum Products	66.2	0.3	5.9	0.2	2.9	0.1
12. Other Manufacturing	63.0	7.3	4.7	0.1	0.5	7.3
Total	55.3	6.8	12.4	1.1	3.9	2.1

Source: The sales of US-owned foreign affiliates in 1975 can be found in US Department of Commerce, *Survey of Current Business*, February 1977. The international production of US firms in 1974 was then estimated by extrapolating the annual growth from 1975–82 back to 1974, sources for 1982 being cited under Table 5.1. The sales of West German-owned foreign affiliates in 1976 were taken from Deutsche Bundesbank, 'Die Kapitalverflechtung der Unternehmen mit dem Ausland nach Ländern und Wirtschaftszweigen 1976 bis 1981', Supplement to *Statistische Beihefte zu den Monatsberichten der Deutschen Bundesbank*, series 3., Zahlungsbilanzstatistik, No. 6, June 1983. The 1974 figures were obtained by applying the 1976–82 annual growth rate over the period 1974–82. The foreign capital stock of UK firms in 1974 is given in Department of Trade and Industry, Business Monitor M4 Supplement, *Census of Overseas Assets, 1974*, and it was assumed that the annual growth of foreign capital stock from 1974 to 1981 mirrored the annual growth of production from 1974 to 1982. For Italy, France and Japan the sources are as cited under table 5.1, and it was assumed that the growth of foreign capital stock from 1974 to 1982 was equal to the growth of international production. Due to major deficiencies in the foreign direct investment data, the Italian estimates were substantially adjusted using the Dunning and Pearce (1985) evidence on the international production of Italian firms over the period 1972–82. The total world international production in 1974 was calculated by a method that relied upon the share of the total accounted for by the firms of the six major industrialised countries in 1982 by industry, and on the change in that share for firms in the Dunning and Pearce sample over the period 1972–82.

Table 5.6   National firms' shares of international production by
manufacturing sector, 1982 (%)

	USA	West Germany	UK	Italy	France	Japan
1. Food Products	31.0	2.5	34.7	0.3	1.8	2.4
2. Chemicals	37.2	27.4	13.4	2.4	4.6	5.8
3. Metals	34.5	10.3	15.8	0.5	5.5	3.3
4. Mechanical Engineering	38.1	13.9	12.5	0.5	2.4	2.9
5. Electrical Equipment	49.3	13.1	12.5	1.3	6.8	11.0
6. Motor Vehicles	40.5	18.1	8.3	4.7	9.4	8.1
7. Other Transport Equipment	25.6	11.1	13.3	0.4	3.5	0.8
8. Textiles	29.8	15.0	42.7	0.7	4.0	8.0
9. Rubber Products	43.5	7.8	21.9	10.5	0.7	6.0
10. Non-metallic Mineral Products	28.6	9.7	22.3	0.1	21.3	6.2
11. Coal and Petroleum Products	53.0	0.5	19.6	0.9	5.4	0.3
12. Other Manufacturing	70.7	8.8	7.3	0.1	1.0	14.9
Total	42.4	10.6	18.3	1.6	5.4	4.9

Source: As for table 5.1.

them to over 50% above their 1974 levels of market share.

A similar story emerges from changes in the shares of total international economic activity (international production and domestic exports combined) as shown in Tables 5.7 and 5.8, though here the gains of West German and Japanese firms are somewhat less marked. The reason is that they had begun to switch the emphasis of their growth from exports to international production rather rapidly at this time. Another factor is that 1982 was an unusually buoyant year for US and UK exports of manufactures by the standards of the 1980s. Once again, motor vehicles, non-metallic mineral products, and coal and petroleum products were major areas of retreat for US firms, but when considering total international economic activity in this way, the mechanical engineering and rubber products sectors also saw a substantial loss of US company shares. In mechanical engineering the major gainers were Italian, French and Japanese firms which all have fields of technological advantage in this area; while in

Table 5.7   National firms' shares of international production plus domestic exports of manufactured goods, 1974 (%)

	USA	West Germany	UK	Italy	France	Japan
1. Food Products	24.4	1.8	10.7	1.6	4.7	1.1
2. Chemicals	30.0	21.0	9.0	4.6	7.6	5.9
3. Metals	17.6	13.5	8.7	2.9	7.2	13.2
4. Mechanical Engineering	39.1	15.3	8.1	2.9	3.9	6.1
5. Electrical Equipment	41.3	14.9	7.2	2.6	5.4	9.9
6. Motor Vehicles	48.2	16.8	7.7	4.5	8.0	11.3
7. Other Transport Equipment	27.2	9.2	9.9	1.5	5.3	23.5
8. Textiles	1.24	11.7	13.9	8.8	7.1	8.9
9. Rubber Products	39.2	10.6	7.0	9.7	8.3	6.8
10. Non-metallic Mineral Products	33.6	11.0	15.8	5.6	10.4	6.4
11. Coal and Petroleum Products	63.2	1.2	6.0	1.1	3.5	0.2
12. Other Manufacturing	28.8	12.0	7.4	2.1	3.1	8.6
Total	35.1	10.4	8.7	3.1	5.6	6.8

Source: International production, as for Table 5.5, and total exports, UN, *Yearbook of International Trade Statistics*, 1976. For the UK the exports of foreign-owned manufacturing affiliates in 1974 are given by Department of Trade and Industry, Business Monitor M4, *Overseas Transactions, 1974*. For West Germany the local production of foreign-owned affiliates in 1976 and 1982 was taken from the sources cited under Tables 5.1 and 5.5, and it was assumed that the annual growth of foreign affiliate production between 1976 and 1982 was equal to the annual growth of foreign affiliate exports from 1974 to 1982. For all other countries it was supposed that the annual growth of inward international production from 1974 to 1982 was equivalent to the annual growth of the stock of inward direct investment from 1975 to 1982 (using sources cited under tables 5.1 and 5.2), and that the propensity to export out of that production was the same in 1974 as it was in 1982.

rubber products the most headway has been made by advantaged British and Japanese companies.

If it is true that the international capital accumulation of firms has been allied to the pattern of their technological accumulation, then the sectoral distribution of their production networks should show the same kind of stability over time that their patenting activity does. The response of European and Japanese firms to the earlier internationalisation of US companies can be expected from the findings of Chapter 4 to have been greatest in their respective

Table 5.8 National firms' shares of international production plus domestic exports of manufactured goods, 1982 (%)

	USA	West Germany	UK	Italy	France	Japan
1. Food Products	24.0	3.6	18.3	1.9	5.3	1.6
2. Chemicals	27.0	23.2	11.9	3.5	7.0	5.6
3. Metals	15.3	12.7	8.5	4.5	7.4	13.0
4. Mechanical Engineering	28.1	17.2	8.4	5.4	5.1	10.0
5. Electrical Equipment	35.5	12.5	8.7	2.9	6.7	19.4
6. Motor Vehicles	25.8	21.3	5.5	4.0	8.4	20.2
7. Other Transport Equipment	28.8	13.7	8.7	2.3	8.2	13.7
8. Textiles	8.7	10.6	9.4	11.8	5.4	7.1
9. Rubber Products	29.5	11.4	15.0	9.3	7.6	9.0
10. Non-metallic Mineral Products	20.3	11.5	15.6	6.5	16.3	8.3
11. Coal and Petroleum Products	46.5	1.1	17.2	2.3	6.0	0.4
12. Other Manufacturing	27.0	11.0	6.4	5.3	3.8	11.1
Total	27.6	12.2	11.3	4.2	6.4	9.4

Source: As for table 5.2

areas of traditional strength, such that their relative strengths and weaknesses in international production are liable to have been reinforced. The stability of the industrial pattern of activity of national groups of companies was tested by examining the relationship between the cross-industry distribution of shares in international production in 1982 and 1974, for the firms of each of the six major industrialised countries. The following cross-section regression was run for each national group of firms:

$$\text{SIP}_{it} = \alpha + \beta\text{SIP}_{it-1} + \varepsilon_i$$

SIP depicts the share of total international production that the firms of the country in question account for, in industry i at time t (where t is calculated for the year 1982 and $t-1$ is represented by 1974). As in the analysis of Chapter 2, the t-statistic on the estimated $\hat{\beta}$ coefficient measures the degree of positive correlation between the two distributions, while the relative magnitudes of $\hat{\beta}$ and the estimated Pearson correlation coefficient $\hat{R}$ determine

Table 5.9   The results of the crosss-industry regression of national
firms' share of international production in 1982
on their share in 1974

Country	$\hat{\alpha}$	$\hat{\beta}$	t_α	t_β	$\hat{R}$
USA	10.806	0.590	1.20	3.41**	0.733
West Germany	4.338	0.730	1.77	3.59**	0.750
UK	9.475	0.547	2.05	2.40*	0.605
Italy	0.375	0.845	1.46	12.35**	0.969
France	0.286	1.229	0.17	4.36**	0.809
Japan	0.595	1.742	0.68	7.35**	0.919

*   Denotes coefficient significantly different from zero at the 5% level.
**   Denotes coefficient significantly different from zero at the 1% level.
Number of observations = 12.

whether the degree of specialisation has risen or fallen. The degree of specialisation in international production rises or becomes narrower where $|\hat{\beta}| > |\hat{R}|$, and it falls or becomes broader where $|\hat{\beta}| < |\hat{R}|$. Each distribution is constructed across only 12 industries, as this is the maximum extent of disaggregation of international production that is feasible.

The results are collected together in Table 5.9. There is a positive association between the industrial pattern of international production in 1982 and 1974 for the firms of all six countries. It is slightly weaker for British firms, for which it is significant at the 5% level, while in all other cases it is significant at the 1% level. The rapid growth of international production in the 1970s and early 1980s has tended to run along the channels set by earlier MNC operations, and so the pattern of these operations has not been greatly disrupted.

The degree of specialisation in international production has fallen for US, British, Italian and to a lesser extent for West German firms. In the case of the more mature MNCs of the USA and UK, the rapid internationalisation of their major competitors seems to have caught them up fastest in those areas in which they were previously the only large firms to serve markets through international production rather than exports. With Italian firms the broadening out of specialisation in international production may reflect the fact that internationalisation has only just begun to gain ground, and so new MNCs are gradually becoming active.

Table 5.10   The results of the cross-industry regression of national
firms' share of international production plus domestic
exports in 1982 on their share in 1974

Country	$\hat{\alpha}$	$\hat{\beta}$	t_α	t_β	$\hat{R}$
USA	6.303	0.594	1.59	5.44**	0.864
West Germany	0.497	1.035	0.32	8.60**	0.939
UK	8.200	0.316	1.84	0.69	0.212
Italy	1.170	0.954	1.57	6.15**	0.889
France	0.389	1.108	0.21	4.00**	0.784
Japan	4.237	0.673	1.72	2.81*	0.664

*   Denotes coefficient significantly different from zero at the 5% level.
**   Denotes coefficient significantly different from zero at the 1% level.
Number of observations − 12.

Meanwhile, the degree of specialisation in international production rose or became narrower for Japanese and French firms. This has happened at a time when international production was growing fastest amongst firms with investments in other industrialised countries. This is linked to the shift away from the traditional commitments of French MNCs in Africa, and away from Japanese investments in particular sector in Southeast Asia, and towards the setting up of production in other developed countries which better reflects the pattern of innovative activity of each national groups of firms. The new type of international production has promoted technological competition far more than the old type did.

The analysis is extended in Table 5.10 by a consideration of trends in the pattern of total international economic activity over time. In this case exports sourced from a domestic production base are included as well as international production. If the share of international production and domestic exports combined held by the firms of a particular country is denoted by $SIPX_{it}$ in industry i at time t, the appropriate simple cross-section regression is given by:

$$SIPX_{it} = \alpha + \beta SIPX_{it-1} + \varepsilon_{it}$$

Once domestic exports are allowed for in this way, the positive correlation between the industrial pattern of market shares of each national group of firms over time becomes stronger for US and West German firms, but weaker for Italian and Japanese companies.

As measured by the t-statistic on the estimated $\hat{\beta}$ coefficient, the correlation remains significant at the 1% level for US, West German, Italian and French firms, but this is only true at the 5% level for Japanese firms. For UK companies there is no longer a significant association between the industrial composition of market shares in 1982 and 1974, but instead a noticeable change in the distribution of activity once exports from the UK are taken into account. This result recalls the finding of Chapter 2 that it is British companies that have experienced the greatest shift in the industrial pattern of their innovative activity in the period leading up to the 1980s. As was suggested earlier, it seems likely that the disruption of the composition of their innovative activity is likely to have been linked to a sizeable change in productive activity, with formerly strong companies cutting back the scale of their operations or going out of business entirely. It is also clear from Tables 5.7 and 5.8 that there have also been some British companies that have been very successful in this period, especially in the food products, rubber products and coal and petroleum sectors.

The degree of specialisation in total international economic activity has in general been rising, except for US firms. This is again in accord with the general increase in the degree of techno-logical specialisation noted in Chapter 2. The major anomaly here is the case of US firms, whose specialisation in innovative activity has become narrower, but whose specialisation in productive activity has been broadened. The explanation probably lies in the process by which European and Japanese firms have caught up with those of the USA. As US companies have slipped back from 35.1 to 27.6% of international economic activity (Tables 5.7 and 5.8), they have increasingly concentrated their technological ef-forts in the fields of their greatest strengths. At the same time, they lost market shares in sectors in which they had historically led the way in the internationalisation of production, as other firms moved in the direction of international networks. The key example here is the motor vehicles sector, in which US companies held a relatively advantaged position in 1974, but an average one by 1982 (relative to the share of other US firms in world markets). An important factor in this case has been the shift away from the scale-intensive technologies of the 1960s towards electronics-related technologies, which particularly in this industry has favoured innovative Japanese and German companies with a different technological base to that of the US firms.

5.4   The Sectoral Pattern of Success and Failure of Countries by Comparison with their Firms

The period between the early 1970s and 1980s was one in which the process of other countries catching up with the export shares of the USA was transformed into one in which the firms of other countries caught up with the total international economic activity of US firms. In fact, the USA was responsible for around 14% of total world exports of manufactured goods in both 1974 and 1982 (see Tables 5.11 and 5.12). Although the industrial competitiveness of the USA may have undergone a further decline since 1982, it held up quite well in 1974–82, in part for the same reason that European exports had risen rapidly in the late 1950s and early 1960s; that is, due to the growth of the local operations of foreign-owned affiliates. The entry and expansion of foreign firms, especially in the electrical equipment sector, appears to have played a beneficial role. This is a sector in which domestic firms have had areas of technological advantage which have enabled them to respond to the demands of international competition.

The relationship between the industrial composition of the national share of exports of manufactures for each country in 1974 and 1982 was tested using a simple cross-industry regression. Denoting the national share of total world exports in industry i at time t by SX_{it}, where t is represented by the year 1982 and $t-1$ by 1974, then the estimated equation may be written:

$$SX_{it} = \alpha + \beta SX_{it-1} + \varepsilon_{it}$$

On the whole, the sectoral distribution of the growth in exports in the 1974–82 period reinforced the existing pattern of national specialisation in manufacturing, just as had happened with the international economic activity of national groups of companies. This is clear from Table 5.13, which shows that there was a significant correlation between the pattern of exports in 1982 and 1974 for every country. It was significant at the 5% level for the UK, and at the 1% level for every other country. Once again, it is the UK which displayed the greatest mobility effect, though not to the same extent as the pattern of activity of its firms. The degree of specialisation in international trade in manufactures rose or became narrower for the USA, Italy and Japan, while it fell or became broader for West Germany, the UK and France.

Table 5.11   National shares of exports by manufacturing sector, 1974 (%)

	USA	West Germany	UK	Italy	France	Japan
1. Food Products	16.7	3.3	2.5	2.5	7.5	0.9
2. Chemicals	14.7	20.1	8.3	5.2	8.2	6.7
3. Metals	7.3	17.2	2.8	3.9	7.5	17.1
4. Mechanical Engineering	23.8	23.1	10.7	5.5	6.8	8.1
5. Electrical Equipment	14.1	15.3	5.4	4.2	5.8	12.0
6. Motor Vehicles	17.2	21.7	6.6	5.2	9.1	15.9
7. Other Transport Equipment	28.1	7.9	5.4	2.0	5.4	25.9
8. Textiles	6.1	12.1	8.0	10.9	8.3	9.1
9. Rubber Products	13.4	18.3	7.2	7.4	13.7	9.0
10. Non-metallic Mineral Products	8.5	18.8	7.5	11.9	10.4	7.8
11. Coal and Petroleum Products	7.9	7.1	4.0	5.4	2.7	0.6
12. Other Manufacturing	13.3	13.6	9.1	4.3	5.8	8.5
Total	14.0	14.0	5.9	4.9	7.1	9.4

Source: United Nations, *Yearbook of International Trade Statistics*, 1976.

When considering national shares of exports of manufactures, the main question of interest in the present context is the relationship between the pattern of success and failure of countries in international competition and the success and failure of their firms. Are locations which are attractive as production sites in a given industry also the homes of strong national firms with a high level of international economic activity? This was tested by means of a cross-industry regression relating national firms' share of international production plus domestic exports in industry i ($SIPX_i$) to the national share of exports (SX_i) at a given point in time, as follows:

$$SIPX_i = \alpha + \beta SX_i + \varepsilon_i$$

The results are reported in Tables 5.14 and 5.15. In 1974, the international economic activity of countries and their firms was positively correlated for West Germany, Italy, France and Japan.

Table 5.12 National shares of exports by manufacturing sector, 1982 (%)

	USA	West Germany	UK	Italy	France	Japan
1. Food Products	16.4	5.3	4.0	3.2	8.3	0.8
2. Chemicals	16.3	17.6	8.9	4.4	9.7	5.2
3. Metals	6.6	14.6	5.3	6.0	7.9	16.4
4. Mechanical Engineering	22.2	19.1	9.3	7.2	6.7	11.8
5. Electrical Equipment	20.1	12.0	6.2	4.2	5.6	22.1
6. Motor Vehicles	11.3	22.4	4.3	3.6	7.7	24.7
7. Other Transport Equipment	28.4	13.8	7.5	3.2	8.6	15.4
8. Textiles	5.2	9.9	4.5	13.4	5.9	6.7
9. Rubber Products	11.1	16.3	7.3	7.3	12.3	11.0
10. Non-metallic Mineral Products	7.9	14.4	5.8	13.6	9.3	9.7
11. Coal and Petroleum Products	5.6	4.7	3.8	4.4	3.0	0.4
12. Other Manufacturing	14.4	11.6	7.3	6.8	5.5	9.4
Total	14.2	13.2	6.1	5.8	7.0	11.0

Source: United Nations, *Yearbook of International Trade Statistics*, 1984.

Table 5.13 The results of the cross-industry regression of the national share of exports in 1982 on the national share in 1974

Country	$\hat{\alpha}$	$\hat{\beta}$	t_α	t_β	$\hat{R}$
USA	−0.525	1.004	−0.25	7.56**	0.922
West Germany	2.462	0.741	1.22	5.91**	0.881
UK	2.968	0.496	2.48*	2.86*	0.671
Italy	0.154	1.104	0.17	7.56**	0.922
France	1.971	0.732	1.65	4.94**	0.842
Japan	2.986	0.803	1.06	3.47**	0.740

* Denotes coefficient significantly different from zero at the 5% level.
** Denotes coefficient significantly different from zero at the 1% level.
Number of observations = 12.

This correlation was strong enough to be significant at the 1% level. By contrast, no such association existed for either the USA or

Table 5.14   The results of the cross-industry regression of national firms' share of international production plus domestic exports on the national share of exports in 1974

Country	$\hat{\alpha}$	$\hat{\beta}$	t_α	t_β	$\hat{R}$
USA	31.494	0.158	3.10**	0.24	0.077
West Germany	0.185	0.766	0.07	4.87**	0.838
UK	8.598	0.106	3.44**	0.29	0.089
Italy	0.098	0.680	0.08	3.39**	0.731
France	1.467	0.623	1.13	3.87**	0.774
Japan	−0.102	0.848	−0.17	16.72**	0.982

*   Denotes coefficient significantly different from zero at the 5% level.
**   Denotes coefficient significantly different from zero at the 1% level.
Number of observations = 12.

Table 5.15   The results of the cross-industry regression of national firms' share of international production plus domestic exports on the national share of exports in 1982

Country	$\hat{\alpha}$	$\hat{\beta}$	t_α	t_β	$\hat{R}$
USA	21.592	0.345	3.51**	0.87	0.265
West Germany	−2.223	1.090	−0.98	6.93**	0.910
UK	15.455	−0.700	3.45**	−1.00	0.303
Italy	0.583	0.680	0.52	4.44**	0.815
France	3.118	0.550	1.05	1.47	0.421
Japan	1.123	0.794	1.99	18.72**	0.986

*   Denotes coefficient significantly different from zero at the 5% level.
**   Denotes coefficient significantly different from zero at the 1% level.
Number of observations = 12.

the UK. The critical factor in explaining the difference between these two groups is the degree of international maturity of their MNCs. The well-established MNCs of the USA and the UK are today far less dependent upon their domestic operations, and the pattern of their success and failure bears far less relationship to that of their home countries. There was a positive correlation between the performance of the USA and the UK and their firms in 1974, but it was not significant.

By 1982, this contrast was if anything stronger. The correlation between the pattern of total international economic activity of firms and the international trade of countries had become stronger for West Germany, Italy and Japan, but had turned into a negative (though still insignificant) association for the UK. However, French firms seem to be becoming rather more like the American and British, with the growth of their international economic activity beginning to take place in areas in which their home economy has not been particularly favoured.

The national divisions are similar when it comes to the degree of specialisation. The firms of West Germany, Italy, France and Japan were less specialised in their activity than their countries were in international trade in 1974. US and UK firms, though, were more specialised than were their home countries. It seems that the more locationally dispersed the activity of firms is, the more they can specialise in the sectors in which they are strongest; or, put another way, it is the strongest groups of firms that have the greatest potential for expansion as they become internationalised. By 1982, French and West German firms displayed this same trait of becoming more specialised in their international economic activity than were their home countries, which again may reflect the progress made by their internationalisation over the 1974–82 period.

5.5   Summary and Conclusions

In one sense the experience of the major industrialised countries and their manufacturing firms between the early 1970s and the early 1980s was the reverse of what it had been between the mid-1950s and mid-1960s. In the latter case, the USA had lost shares of world trade in manufactures, but US firms had continued to expand rapidly and to lead the firms of other countries once their international production is taken into account. By the late 1960s, European and Japanese firms had also begun to catch up with their US rivals, and especially in areas of traditional strength their domestic exports steadily rose. At this stage, they began to embark upon an increased internationalisation of their own. From the early 1970s to the early 1980s the USA held on to its share of world trade in manufactures, but US manufacturing companies lost market shares quite heavily.

The US share of exports of manufactures even rose slightly between 1974 and 1982, from 14.0 to 14.2%. US firms, however, saw their share of total world markets for manufactures served by exports or international production slip from 35.1 to 27.6%. European and Japanese companies made substantial gains at their expense. The firms of all the other five major industrialised countries all increased their market shares. Particularly noticeable, perhaps, were the gains made by UK firms, from 8.7 to 11.3% of international production and domestic exports of manufactured goods combined. This stands out since the advances of British companies came at a time when Britain at best held a static position as a trading nation in manufacturing. British MNCs have done especially well through increasing international production. Despite the already greater maturity of their international networks, the absolute increase in their share of international production, from 12.4 to 18.3%, was greater than any other national group of firms, and the proportional rate of increase was on a par with that of French and Italian firms. It was still, however, behind the very high proportional rate of international growth achieved by the West Germans and the Japanese.

There remains an important difference, though, between the type of expansion experienced by British firms, and that which has characterised the firms of the other European countries and Japan. This is that UK firms have changed the industrial pattern of their activity significantly in the course of their recent growth. This has been particularly evident in the composition of their total international economic activity including exports from the UK, even if their international production has held to a more stable pattern. This may be related to the greater disruption that the pattern of their innovative activity has undergone, as noted in Chapter 2. The growth of the firms of other countries has more or less retained the established industrial distribution of activity, and has in this respect taken place along rather more predictable lines. This is as true of US firms who have suffered a retrenchment in their market shares, as it is of other European and Japanese companies whose shares have risen.

British and American firms are more similar and stand out together when it comes to their high degree of international maturity. The consequence of their earlier internationalisation is that the connection between the locational strengths of their home country and their own advantages in international economic activity has been severely weakened. This association still persists for Japanese, West German, Italian and to a lesser extent for French firms.

What until now has only been discussed in a descriptive rather than a statistical way is the relationship between the success and failure of firms and the structure of their innovative activity or technological advantages. A variety of evidence in the preceding two or three chapters has pointed to some kind of link between the cross-industry distribution of innovation as analysed in Chapter 2, and the distribution of international economic activity controlled by the firms of each of the major industrialised countries. A more formal appraisal of how productive activity has depended upon the prevailing pattern of technological accumulation is to be found in Chapter 6.

6

Technological Advantage as a Determinant of the International Economic Activity of Firms

6.1   Introduction

This chapter considers empirical evidence on the relationship between technological advantage and the total international economic activity of firms from the six major industrialised countries in the early 1980s. Previous work on innovation and international trade and production has rarely considered the significance of technological advantage in the competitive success of firms of particular countries. The neo-technology theories of international trade (as surveyed, for example, by Soete, 1981) have tended to treat technology as an additional domestic factor of production, supportive of a country's trading position in high-technology intensive goods. Since they have not focused on the dynamic role of technological innovation in generating the advantages which enable a country's firms to expand, they have tended to neglect international production and other forms of servicing international markets. Where technology gap trade theories have been extended to allow for different modes of international activity, as in the case of the product cycle model, they have only been concerned with a certain kind of import-substituting international production. They then attempt to explain the circumstances under which the location of production might shift, that is that firms switch from exporting to international production.

As it has recently developed, the theory of international production (as set out, for example, in Dunning, 1981, or Caves, 1982) has typically addressed a related set of issues. It has, on the one hand, attempted to answer questions about the geographical location of production by MNCs, the export versus local production decision, and its effect on the industrial structure of national economies (see, for example, Hirsch, 1976; Caves, 1980; and Dunning, ed., 1985). On the other hand, theoretical work in the Coasian tradition has been concerned with why MNCs in general have grown relative to intermediate product trade through external markets, one aspect of which is the international production versus licensing decision (see, for example, Buckley and Casson, 1976; Casson; 1986, 1987; and Casson et al., 1986). Little attention has been paid to the issue of why it is that, within any given industry, some firms rather than others have become the most successful MNCs, or why Japanese companies have been growing faster than their international rivals. While discussions of this kind can be found in the literature of business organisation and strategy (see, for example, Doz, 1986)[1], they are rarely reflected in treatments that derive from economic theory.

What has been largely lost from the recent theory of international trade and production is a notion of an active firm setting about the generation of ownership advantages (in part through technological accumulation), and thereby expanding its total international involvement. In most recent literature ownership advantages are relegated to, at best, a secondary role. The firm is typically seen as a passive reactor, whose locational decisions are uniquely determined by changes in the international pattern of demand, costs, and government interventions; and whose decisions on expansion are determined by changes in the relative costs of organising transactions through external markets as opposed to the costs of administrative coordination. For this reason, empirical investigations of MNC activity by economists most often deal with issues related to the modality of serving international markets (by exports, international production or licensing and other contractual agreements). They generally leave aside the question of why it is, within such markets, that some firms are growing relative to others.

The prevalent procedure is particularly regrettable where exports, international production and licensing are complementary with, rather than substitutable for, one another as means of expansion. This may well be the case in those modern industries characterised by a high degree of intra-industry trade and production, especially where this is associated with technological competition between the leading international companies. The most advantaged firms may engage in both exporting and direct production in the markets of their competitors, as well as establishing selective cross-licensing agreements with these same major rivals.

Moreover, the relative expansion of total international involvement by the strongest firms, compared with the position of less capable firms, cannot easily be divorced from the relative importance of international trade, production and licensing. This has been recognised in the literature on the use of joint ventures and contractual arrangements by MNCs (see, for example, Cantwell and Dunning, 1985). Firms whose production is characterised by more mature and standardised technology and which are less internationally experienced, are more likely to enter markets through collaborative agreements with other firms. Hence, industries which are dominated by a small group of technology leaders are likely to have a different pattern of international trade, production and licensing than industries in which an extensive competitive fringe prevails. This chapter therefore marks something of a departure from the orthodox literature on international trade and production by directly confronting the issue of the determinants of competitive success. It investigates the extent to which the industrial pattern of the total international economic activity of firms is dependent upon the sectoral distribution of their technological advantage.

Section 6.2 proceeds by calculating an index of revealed comparative advantage (RCA) for each of the six countries, and for the nationally owned firms of these countries. Each index is for a 12-industry distribution, and is based on the data described in Chapter 5. Section 6.3 presents the results of statistical tests using these indices. Firstly, it tests the hypothesis that, for each country, the RCA of its firms is positively correlated with their index of revealed technological advantage (RTA). Secondly, for each country, it is hypothesised that in the earliest stages of international growth (technological and capital accumulation) of their firms, export success and international production are complementary and both positively related to technological advantage, but that as internationalisation proceeds location advantages become more important in determining a balance between exports and international production. A positive correlation between the RCA of a country and the RTA of its firms is consequently expected only where that country's firms are at an early stage of internationalisation. This follows from the result of the previous chapter that the shares of some countries in world exports are linked to the shares of their firms in total international economic activity, but not where mature MNCs predominate amongst the country's firms, as has happened amongst US and UK companies.

6.2   Industrial Patterns in the Economic Advantages of Countries and their Firms

The data on international trade and production developed in the previous chapter were used to generate indices of comparative advantage in the economic activity of countries and their firms. Following Balassa (1965), the revealed comparative advantage of countries in international trade (RCAC) is defined as follows:

$$RCAC_{ij} = (X_{ij}/\Sigma_j X_{ij})/(\Sigma_i X_{ij}/\Sigma_i\Sigma_j X_{ij})$$

The revealed comparative advantage of industry i for country j is denoted by $RCAC_{ij}$, while the value of exports of industry i from country j is denoted by X_{ij}. Where the index takes a value greater than one the industry is advantaged, while where it is less than one the sector is disadvantaged for the country in question.

Now the revealed comparative advantage of each country's firms in international economic activity (RCAF) is given as follows:

$$RCAF_{ij} = [(IX_{ij} + IP_{ij})/\Sigma_j(IX_{ij} + IP_{ij})]/[\Sigma_i(IX_{ij} + IP_{ij})/ \Sigma_i\Sigma_j(IX_{ij} + IP_{ij})]$$

In this case $RCAF_{ij}$ represents the revealed comparative advantage of country j's firms in industry i, IX_{ij} denotes the value of exports by indigenous firms in industry i from country j, and IP_{ij} is the value of international production in industry i controlled by MNCs based in country j. Indigenous firm exports (IX_{ij}) are equal to total exports (X_{ij}) minus the exports of foreign-owned affiliates in host country j. The distribution of advantaged and disadvantaged industries in the index again varies around the value of unity.

Using patent data as discussed in greater detail in Chapter 2, but here using a 12-industry distribution, an index of revealed technological advantage was calculated according to the formula:

$$RTA_{ij} = (P_{ij}/\Sigma_j P_{ij})/(\Sigma_i P_{ij}/\Sigma_i \Sigma_j P_{ij})$$

P_{ij} represents the number of US patents in industry i granted to residents of country j. In the case of the USA, aggregation is carried out over all US patents, while for all other countries aggregation over j excludes patents granted to US residents, and so the RTA index reflects advantage relative to all non-US firms. The reason is to focus on foreign patenting as far as possible, even though this is thought to be correlated with inter-industry patterns of domestic patenting. The RTA index is drawn up in such a way that it is normalised for systematic inter-industry and inter-country differences in the propensity to patent.

The RTA index for a 12-industry distribution was calculated for the period 1972–82, as shown in Table 6.1. The RCAC and RCAF indices were calculated for the year 1982, but there are good reasons for assessing RTA over a span of years leading up to this. The main reason is that even though there is a lag between the application for and the granting of a patent, it is likely to take some time before a technical advantage is properly reflected in the market shares of a company. However, in general, the span of years chosen does not greatly affect the analysis, since the sectoral pattern of technological advantage has remained relatively stable over this time (see Chapter 2).

It can be shown that the extent of dispersion of the RTA index around the mean is lower in the case of the US than for any other country. In part, this reflects the different way of calculating the index for the US, and the high share of US patents attributable to US residents (nearly 64% over the period 1972–82). Yet it also partly represents a lower degree of technological specialisation amongst US firms, who hold advantages in a broader range of industries than the firms of other countries. However, according to Table 6.1, US firms are particularly advantaged in coal and petroleum products, food products, metals, electrical equipment, aerospace, rubber products, and other manufacturing (paper and wood products).

West German firms hold a position of technological advantage

Table 6.1   The revealed technological advantage index for a
12-industry distribution, 1972–1982

	USA	West Germany	UK	France	Italy	Japan
1. Food Products	1.09	0.60	1.12	0.83	0.70	0.96
2. Chemicals	0.91	1.17	1.04	1.04	1.29	0.89
3. Metals	1.08	0.89	1.02	1.08	0.85	0.81
4. Mechanical Engineering	0.97	1.10	0.96	0.96	1.18	0.79
5. Electrical Equipment	1.01	0.85	0.98	1.09	0.82	1.30
6. Motor Vehicles	0.93	1.19	0.99	1.03	0.82	1.10
7. Other Transport Equipment	1.01	1.07	1.27	1.31	0.85	0.88
8. Textiles	0.92	1.23	1.24	0.92	0.79	0.94
9. Rubber Products	1.01	1.06	1.09	0.96	1.06	1.11
10. Non-metallic Mineral Products	0.99	0.85	1.42	1.06	0.71	1.00
11. Coal and Petroleum Products	1.32	0.64	1.36	1.42	0.66	0.72
12. Other Manufacturing	1.05	0.85	0.88	0.85	·0.77	1.21

Source: US Patent and Trademark Office, Office of Technology Assessment
and Forecast, unpublished data on US patent counts.

in textiles, motor vehicles, chemicals, mechanical engineering,
other transport equipment, and rubber products. This helps to
explain how West German firms proved capable of a vigorous
response to the competitive challenge of US direct investment in
the 1960s in the motor vehicles and chemicals sectors (as documented in Chapter 4).

UK firms are technologically advantaged in food products and
rubber products, in which Chapter 4 suggested they had regained
market shares from US competitors, and also in chemicals and
metals, sectors in which some UK firm revival was suggested by
Chapter 3. The RTA index for the UK is also substantially greater
than unity in the case of non-metallic mineral products, coal and
petroleum products, other transport equipment (aircraft), and
textiles.

French firms seem to have recaptured market shares from the
French affiliates of US firms in the early 1970s in the metals and
electrical equipment sectors, and in both these they show a technological advantage. According to Table 6.1, French firms also
hold a technological advantage in coal and petroleum products,
other transport equipment, non-metallic mineral products, chemicals, and motor vehicles.

The position of Italian firms is less clear, since in sectors where
they were apparently able to show some response to the American

challenge of the 1960s, they are not necessarily technologically advantaged. Indigenous Italian firms seem to have recovered ground in the 1970s in the mechanical engineering, electrical equipment and metals sectors, and yet of these it is only in the mechanical engineering industry that the RTA index takes a high value. Italian firms are, on the other hand, technologically advantaged in chemicals and rubber products. The reason for the discrepancy may be that technological accumulation is not always properly reflected in patenting in the USA in the case of Italian firms (Archibugi, 1986).

By way of contrast, the technological advantages of Japanese firms lie in electrical equipment, other manufacturing (notably professional and scientific instruments), rubber products, and motor vehicles. It is in these sectors that most is heard of the Japanese challenge, which has nowadays replaced the American challenge, and affected the position of US as well as European firms.

The index of revealed comparative advantage of countries (RCAC), over the same 12-industry distribution in 1982, is set out in Table 6.2. The contrast with Table 6.1 is interesting. For example, the USA as a country has a comparative advantage in mechanical engineering and chemicals, in which its firms have not particularly strong technological advantage. However, US firms

Table 6.2   The index of revealed comparative advantage of countries across a 12-industry distribution, 1982

	USA	West Germany	UK	France	Italy	Japan
1. Food products	1.15	0.40	0.66	1.18	0.56	0.08
2. Chemicals	1.15	1.34	1.46	1.39	0.76	0.47
3. Metals	0.46	1.11	0.87	1.13	1.04	1.50
4. Mechanical Engineering	1.57	1.45	1.52	0.96	1.26	1.07
5. Electrical Equipment	1.42	0.91	1.02	0.80	0.74	2.01
6. Motor Vehicles	0.80	1.70	0.70	1.10	0.62	2.26
7. Other Transport Equipment	2.00	1.05	1.22	1.23	0.56	1.40
8. Textiles	0.37	0.75	0.74	0.85	2.32	0.62
9. Rubber Products	0.78	1.24	1.19	1.76	1.27	1.01
10. Non-metallic Mineral Products	0.56	1.09	0.94	1.33	2.36	0.88
11. Coal and Petroleum Products	0.40	0.35	0.62	0.43	0.77	0.03
12. Other Manufacturing	1.01	0.88	1.19	0.79	1.19	0.86

Source: United Nations, *Yearbook of International Trade Statistics*, 1984.

do have a technological advantage in coal and petroleum products and metals, in which the USA as a trading nation is comparatively disadvantaged.

For West Germany, the contrast is greatest in the case of textiles (a strong RTA, but weak RCAC), and metals (a low RTA with a higher RCAC). The UK has a high RTA but low RCAC in the coal and petroleum products, textiles, and food products sectors. A strong RCAC but a weak RTA is observed for France in food products, for Italy in textiles and non-metallic mineral products, and for Japan in metals and other transport.

Table 6.3 shows the values taken by the index of revealed comparative advantage of firms (RCAF) in 1982. Again, some interesting anomalies come to light. Despite the technological advantage of US firms, and the comparative trading advantage of the US as a nation in food products, US companies appear to be disadvantaged in this sector when the total international economic activity of firms is considered. This means that US food firms, while strong at home, perhaps partly because of this, are less internationally oriented than their major competitors from other industrialised countries (most notably the UK).

A general statistical analysis of the relationship between the RCAC, RCAF and RTA indices follows in the next section.

Table 6.3   The index of revealed comparative advantage of each country's firms, across a 12-industry distribution, 1982

	USA	West Germany	UK	France	Italy	Japan
1. Food Products	0.87	0.30	1.62	0.83	0.45	0.17
2. Chemicals	0.98	1.90	1.05	1.08	0.85	0.60
3. Metals	0.55	1.04	0.76	1.15	1.07	1.39
4. Mechanical Engineering	1.02	1.41	0.74	0.79	1.29	1.07
5. Electrical Equipment	1.29	1.02	0.77	1.04	0.70	2.07
6. Motor Vehicles	0.93	1.74	0.49	1.31	0.97	2.16
7. Other Transport Equipment	1.04	1.12	0.77	1.27	0.55	1.46
8. Textiles	0.31	0.86	0.84	0.84	2.84	0.76
9. Rubber Products	1.07	0.93	1.33	1.18	2.24	0.96
10. Non-metallic Mineral Products	0.74	0.94	1.39	2.52	1.57	0.90
11. Coal and Petroleum Products	1.68	0.09	1.53	0.93	0.54	0.04
12. Other Manufacturing	0.98	0.90	0.57	0.59	1.27	1.18

Source: As for table 5.2.

6.3   *The Importance of Technological Advantage in the International Economic Activity of Firms*

If the RTA index is a measure of the industrial distribution of technological advantages amongst the firms of a given country, the RCAC index is a measure of the sectoral distribution of location advantages of the country concerned as a production site for serving world markets. The greater is the extent of the internationalisation of industry, and the more global that firms become in their operations, the less correlation is to be expected between the two. The degree of this correlation can be tested by fitting the following simple cross-section regression for each country:

$$RCAC_i = \alpha + \beta RTA_i + \varepsilon_i$$

An alternative to the hypothesis stated above is suggested by technology gap theories of trade. Although they postulate a positive association between the absolute level of domestic innovative activity and export performance, they might lead us to expect a positive coefficient in this equation expressed in terms of relativities. In Table 6.4 it can be seen that, taking $RCAC_i$ for 1982 and RTA_i for 1972–82, a positive correlation is obtained only in the cases of West Germany and Japan. This is because, in their post-war expansion, West German and Japanese firms have relied more heavily on a strategy of export growth than their major

Table 6.4   The results of the regression of the RCA index of countries in 1982 on their RTA in 1972–1982

Country	$\hat{\alpha}$	$\hat{\beta}$	t_α	t_β	$\hat{R}$
USA	2.222	−1.221	1.53	−0.86	0.263
West Germany	−0.331	1.412	−0.85	3.55**	0.746
UK	1.803	−0.711	3.07*	−1.36	0.396
Italy	1.439	−0.364	1.61	−0.36	0.114
France	1.697	−0.591	2.66*	−0.98	0.297
Japan	−0.808	1.869	−0.74	1.70	0.474

*   Denotes coefficient significantly different from zero at the 5% level.
**   Denotes coefficient significantly different from zero at the 1% level.
Number of observations = 12.

international competitors. Partly due to wartime expropriations, they have historically had less extensive international networks than their American and British rivals.

In fact, it is only in the West German equation that $\hat{\beta}$ is seen to be significantly different from zero. In the other four countries there is actually a negative relationship between $RCAC_i$ and RTA_i. This suggests not only that R&D and innovative activity tends to be more centralised than the overall production of firms, but that relatively more innovative firms tend to become more geographically dispersed in their production activity. For this reason, West Germany and Japan may well become more like the other major industrialised countries in due course. The lack of a positive relationship between $RCAC_i$ and RTA_i is as expected in the US and UK cases, given the international maturity already achieved by their firms. For France and Italy, it is presumably explained by the relatively weaker role of fundamental research and technology of a patented kind in fuelling national export growth.

The sectoral pattern of the competitive success of firms themselves rather than their countries can be measured by the RCAF index. In earlier sections it has been argued that technological advantage is an important means of sustaining a competitive position in international markets. It seems to have been significant in determining the ability of firms from at least some European countries to respond to an increase in technological competition in Europe, generated by the entry of US-owned foreign affiliates after the formation of the EEC in 1958. If this has been carried through to the overall international economic activity of these firms then a positive correlation is to be expected when fitting a regression line of the form:

$$RCAF_i = \alpha + \beta RTA_i + \varepsilon_i$$

The results of this regression, with $RCAF_i$ for the year 1982, and RTA_i for 1972–82, are set out in Table 6.5. A positive correlation is indeed obtained for all six countries, though it is not significant for firms of Italian or French origin. This suggests again that it is firms based in the USA, West Germany, the UK and Japan that are the most embroiled in technological competition with each other, while French and Italian firms are rather more dependent upon traditional organisational, managerial and marketing skills. This

Table 6.5   The results of the regression of the RCA index of each country's firms' production for international markets in 1982 on their RTA in 1972–1982

Country	$\hat{\alpha}$	$\hat{\beta}$	t_{α}	t_{β}	$\hat{R}$
USA	−0.974	1.883	−1.21	2.40*	0.605
West Germany	−0.846	1.949	−1.84	4.15**	0.796
UK	−0.538	1.370	−0.85	2.43*	0.609
Italy	0.794	0.459	0.77	0.40	0.126
France	0.467	0.631	0.51	0.72	0.224
Japan	−0.961	2.076	−1.00	2.13*	0.559

*   Denotes coefficient significantly different from zero at the 5% level.
**   Denotes coefficient significantly different from zero at the 1% level.
Number of observations = 12.

particular division between national groups of firms has only emerged clearly in the past decade. The same regression was performed with RCAF$_i$ for 1974 and RTA$_i$ in 1963–74, from which it emerges that the relationship between RCAF and RTA has become stronger for US, British and Japanese firms, but weaker for those of France and Italy. Little difference was found in the case of West German firms.

The increasing weakness of the correlation between RCAF and RTA amongst Italian companies accords with the apparent observation of Section 6.2 that where Italian firms had recaptured market shares from US-owned foreign affiliates between the mid-1960s and the mid-1970s, they had not necessarily been reliant on technological advantage in order to do so. It is in the case of Italy that the estimated correlation coefficient ($\hat{R}$) is weakest. An explanation of this which concurs with the findings of Acocella (ed., 1985) in an investigation of Italian MNCs is that their innovation is based more on organisational change and the development of local skills than on science or research. Partly because of this, it seems that US patent data are not an ideal measure of technological advantage in the case of Italian firms (see the discussion of Archibugi, 1986).

The case of France is rather more complicated. French firms seem to have had the greatest competitive success *vis-à-vis* US-owned foreign affiliates in the metals and electrical equipment sectors. In these same industries French firms held a technological advantage, and have also held a comparative advantage in overall economic activity (the RCAF index). Indeed, for the first eight industries listed in Tables 6.1 and 6.3 the values of the RTA and RCAF indices seem reasonably close for France. The correlation between the two indices is, however, weakened by differences in industries 9 to 11. The greatest discrepancy occurs in non-metallic mineral products, where a clear but small technological advantage is associated with a massive comparative advantage in the overall economic activity of firms (an $RCAF_i$ value of 2.52). Meanwhile, a strong technological advantage in coal and petroleum products is not matched by any comparative advantage in serving inter-national markets. The strong influence of these two outlying observations helps to ensure a weak correlation coefficient in the French regression.

In general, technological advantage seems to be a necessary but not sufficient condition for competitive success. This certainly applies for US, West German and UK firms. Wherever a value of 1.05 or more is recorded for the RCAF index this is associated with a position of technological advantage. There are, however, cases where technologically advantaged firms have failed to achieve the same degree of overall competitive success (the posi-tion of US food companies, and West German and UK textiles companies).

Leaving aside Italy, there appear to be exceptions to this rule when considering French and Japanese firms. French firms have a comparative advantage in rubber products, but narrowly miss having a technological advantage in this sector. Japanese firms hold a strong position in the metals and other transport equipment sectors, apparently without any technological advantage. This is in part a problem of sectoral aggregation, and partly a problem of the measurement of technological advantage within sectors. Japanese firms are especially strong in ferrous metals, rather than non-ferrous metals or fabricated metal products, and in ferrous metals considered separately they do have a technological advan-tage. In other transport equipment, unlike US or French firms whose strength lies in aircraft, Japanese firms have specialised in shipbuilding. Here, while they lack a high value of the RTA index, this may well be a sector in which technological know-how and local skills are not well reflected in foreign patenting.

Note that the RCAF index is considerably more dispersed (has a higher variance) than the RTA index for the firms of all six countries. This suggests that a relatively small degree of technological specialisation reflects itself in a much greater degree of specialisation in productive activity. It may be that firms that are slightly more technologically capable within their industries than other firms from the same country, tend to be very much more competitively successful. However, another contributory factor is that firms tend to have a higher degree of technological diversification than output diversification (Pavitt Robson and Townsend, 1987b). That is, individual firms draw on a broader range of technological innovation in order to support a relatively narrower specialisation in output produced.

Now it was argued above that a correlation (or lack of correlation) between the RCAC and RTA indices reflected the degree of maturity of the expansion of international networks of a country's firms; while it was expected that RCAF would be correlated with RTA irrespective of the degree of internationalisation, provided that the country's firms are strongly reliant upon basic research capacity, and not primarily on organisational skills. The implication is firstly that a positive correlation between RCAF and RTA (total productive and technological activity) should be a general phenomenon. However, the second implication is that the sectoral distribution of innovation will only explain the industrial pattern of trade (or international production) for groups of companies at

Table 6.6   The results of the regression of the RCA index of each country's firms' international production in 1982 on their RTA in 1972–1982

Country	$\hat{\alpha}$	$\hat{\beta}$	t_α	t_β	$\hat{R}$
USA	−0.067	0.989	0.08	1.23	0.363
West Germany	−1.314	2.501	−2.20*	4.11**	0.792
UK	−0.859	1.687	−0.82	1.82	0.499
Italy	−1.956	3.601	−0.76	1.25	0.367
France	0.303	0.823	0.16	0.45	0.141
Japan	−2.924	4.216	−3.44**	4.91**	0.841

*   Denotes coefficient significantly different from zero at the 5% level.
**   Denotes coefficient significantly different from zero at the 1% level.
**   Number of observations = 12.

an early stage of internationalisation. An alternative hypothesis might be that the RCAF index shows a higher positive correlation with RTA than does the RCAC index (the value of the t-statistic on $\hat{\beta}$ is higher for all six countries), because technological advantage is a good indicator of comparative advantage in international production as opposed to trade. To test this alternative hypothesis an index of revealed comparative advantage in international production (RCAIP) was created as follows:

$$RCAIP_{ij} = (IP_{ij}/\Sigma_j IP_{ij})/(\Sigma_i IP_{ij}/\Sigma_i \Sigma_j IP_{ij})$$

As defined above, IP_{ij} denotes the international production of country j's firms in industry i. For each country, this index for 1982 was regressed on the country's RTA index for 1972–82 in the linear functional form

$$RCAIP_i = \alpha + \beta RTA_i + \varepsilon_i$$

The results are shown in Table 6.6. The correlation is indeed more strongly positive for all countries than that found in the regressions of the trade measure (the RCAC index). However, $\hat{\beta}$ is found to be significantly different from zero only in the cases of West Germany and Japan, precisely the countries whose comparative advantage in trade is also best explained by their RTA index. As suggested above, for West German and Japanese firms at a relatively early stage in the construction of international networks, exports and international production grow together in line with technological advantage, as depicted by the model of Chapter 3. Meanwhile, for US and UK firms, the explanatory power of the RTA index in the RCAF equation does not arise from a still better correlation between RCAIP and RTA. It is the combination of international trade and production to provide the total activity of the firms concerned (having eliminated the exports of foreign-owned firms from the trade component) which ensures the better fit of the regressions reported in Table 6.5, when compared with those in Table 6.4.

The conclusion must be that just as it is misleading to consider trade without international production in the manufacturing industries of the 1980s, so it is also misleading to consider international production without trade. For example, a very poor performance of firms in international trade (owing, say, to weak location advantages for production in their domestic economy), combined with a very strong performance in international production, may result in a simply average performance in overall economic activity. A mediocre value of the RTA index would pick up neither extreme variation of performance in international trade nor production considered separately. This type of locational polarisation is more likely the greater is the maturity of internationalisation of a country's firms.

The fact that the estimated coefficient $\hat{\beta}$ is not significantly different from zero in the US and the UK equations in Table 6.6 is therefore a sign of the maturity of the internationalisation of their firms. Location advantages of production at home and abroad have come to generate substantial divergences between industries in the ratio of international production to trade, which have little to do with the relative strengths of the ownership advantages of firms across industries. In Japan and West Germany, where the internationalisation of firms is still at a relatively young stage, it tends to have been technologically advantaged firms that have taken the lead in both international trade, and now production.

To carry this further, the alternative hypothesis might have been extended to encompass the argument that the ratio of international production to exports is likely to be higher in research-intensive industries, owing to a greater propensity to internalise transactions which involve elements of technology transfer across national boundaries. An inspection of Tables 5.3 and 5.4 above would suggest that this was a rather dubious generalisation, since the ratio was, for the world as a whole, high in the case of non-metallic mineral products, and low in the case of aircraft (the reverse of their rankings in an index of research intensity). However, to test a version of this hypothesis for the firms of individual countries, having allowed for the general determinants of industrial variation in the ratio of international production to exports, it is proposed to examine whether variation in the strength of local innovative activity adds anything to the explanation. A regression of the following form was therefore run:

$$(IP_i/IX_i) = \alpha + \beta(\Sigma_i IP_i/\Sigma_i IX_i) + \gamma RTA_i + \varepsilon_i$$

As before, IP represents international production, and IX indigen-

Table 6.7   The results of the regression of each country's ratio of international production to indigenous firm exports in 1982 on the world ratio and on the country's RTA index in 1972–1982

Country	$\hat{\alpha}$	$\hat{\beta}$	$\hat{\gamma}$	t_α	t_β	t_γ	R^2
USA	−4.222	3.488	3.901	−1.20	11.09**	1.06	0.976
West Germany	−0.137	0.147	0.625	−0.16	1.12	0.78	0.128
UK	−4.044	3.299	3.887	−0.95	6.20**	0.97	0.869
Italy	−0.733	0.146	1.071	−0.92	1.13	1.28	0.203
France	0.734	0.725	−0.471	0.36	2.55 *	−0.22	0.519
Japan	0.296	0.325	0.032	0.22	1.86	0.02	0.305

*   Denotes coefficient significantly different from zero at the 5% level.
**   Denotes coefficient significantly different from zero at the 1% level.
Number of observations = 12.

ous firm exports, for the year 1982. The results are reported in Table 6.7. The general features that apparently give rise to a higher ratio of international production to exports are particularly relevant in the case of the USA, the UK, and France (which have a value of $\hat{\beta}$ significantly different from zero). This again, presumably, reflects the relatively greater maturity of the internationalisation of firms of these countries (especially the USA and the UK). However, the RTA index adds very little to this explanation, as the γ coefficient is not significantly different from zero for any country.

This helps to confirm the impression that, in so far as it is important, technological advantage is supportive of an overall competitive strength in total international economic activity. At an early stage in the international economic growth of firms, export success may be associated with a greater volume of international production (as has happened recently for West Germany and Japan), but as internationalisation proceeds the comparative strength of international trade and production comes to depend on locational factors, which can be expected to vary between countries.

6.4   Conclusions and some Further Suggestions

The central conclusion is that in general technological advantage is a necessary though not sufficient condition for competitive success amongst the largest most advanced firms of the major industrialised countries (principally, the largest MNCs). Using data on patenting activity, this emerges most clearly in the case of those groups of firms which are in the forefront of international technological competition. In future research, it would be useful if the analysis above could be extended to incorporate the share of international markets serviced by means of non-affiliate licensing, and if the degree of industrial disaggregation could be improved to consider distributions across more than 12 industries. The rather broad metals, mechanical engineering, electrical equipment, other transport equipment, and other manufacturing categories are particularly unsatisfactory. The difficulty lies in the further refinement of the data on international production.

It has been shown that West German and British firms, and to a lesser degree French firms, relied on existing technological strengths in reviving their position in international markets in the post-war period. Technological advantage has also been a significant factor in the competitive success of US and Japanese firms. Outside these five major industrialised countries firms from smaller or less developed countries are likely to be less reliant on the same kind of strong advantages in research and patenting activity. Italy is perhaps as close to some of these other countries in this respect as it is to the USA or Japan. It is firms from the major industrialised countries (the strongest firms in world markets), that most need technological advantages linked to fundamental research activity to sustain a high rate of growth, and it is these firms that are most embroiled in technological competition which has now moved to a world scale. This is consistent with the suggestion of Pavitt, Robson and Townsend (1987a) that it is the largest firms that are most dependent upon research and scientific activity, the fruits of which are more easily internalised and incorporated in technological advantage than is the case for production engineering. Smaller firms are more dependent upon the latter in their ability to innovate.

Moreover, for US, Japanese, British and West German firms the role of research in supporting overall economic activity has become more important since the early 1970s. At first glance, this may seem a surprising result, since long wave theorists suggest that there has been a fall in the overall rate of innovation, and its industrial impact, as measured by a slowdown in patenting and a slowdown in rates of productivity growth in the industrialised world. With reference to the competitive strategy of MNCs, this has been associated with a rise in rationalised investment, and attempts to gain advantage from a more carefully integrated international division of labour rather than fresh technological innovation. However, despite this shift in emphasis technological competition has remained at least as important as before.

The main reason is that, despite a fall in the overall rate of industrial growth in the developed countries, the international networks of non-US manufacturing firms have been growing very fast (as set out in Chapter 5). The greater internationalisation of production, and the greater technological diversification of MNCs in recent years, have served to increase the extent of technological competition between the world's largest industrial firms. In the setting up and extension of these international networks, and in the continued competition between them, firms have relied upon technological accumulation at an international level. The consolidation of innovative strengths as the basis for world market growth has figured especially prominently amongst US, Japanese, British and West German firms.

Another factor in the greater significance of technological advantages for these firms by 1982 may be that since the cost and difficulty of research has increased since the mid-1970s (the number of patents has fallen while R&D expenditure has risen) smaller less international firms have been compelled to retreat from research-intensive activities, and so larger stronger MNCs that could afford the cost have been better able to exploit their remaining research achievements. If the newer technologies currently being pioneered in microelectronics and genetic engineering eventually come to stimulate a fresh wave of innovation throughout industry, then technological competition could again become more widely dispersed across the firms of the industrialised countries as it seems to have been in the 1960s. Failing this, an increase in collaboration, joint ventures and cross-licensing arrangements may well be observed amongst the most technologically advanced firms.

Even in these circumstances, technological cooperation is likely to run alongside technological competition rather than substituting for it. Theories which simply counterpose domestic exports, international production and non-affiliate licensing as alternatives are not very helpful to an understanding of the current evolution of international industries. Firstly, for the newer MNCs of West Germany and Japan, domestic exports and international production have been found to be complementary means of growth. Secondly, the rapid expansion of international production by non-US firms has helped to sustain technological competition, but at the same time created fresh scope for cross-licensing agreements between firms. The locational and industrial spread of MNCs in the same sector has tended to become more similar, leading to a greater competitive interaction between them. One of the clearest signs of this interaction has been a rise in intra-industry production across manufacturing sectors, and the role of technological competition in this respect is discussed in Chapter 7.

NOTE

1 The importance of technological advantage in the competitive strategy of major MNCs has also been emphasised in the results of a question-naire survey reported by Wyatt et al. (1985).

Technological Competition and Intra-industry Production in the Industrialised World

7.1 Introduction

In Chapter 2 it was argued that the industrial distribution of innovation amongst the firms of a given country tends to proceed in a cumulative way. Chapters 3 and 4 used this idea to suggest that the European response to the 'American Challenge' had been strongest in areas in which European firms were comparatively technologically advantaged. Chapters 5 and 6 have shown that particularly since the early 1970s the expansion of firms in their advantaged areas has meant the growth of outward international production as well as exports. In the framework developed in Chapter 3 it was further suggested that this trend creates an expectation that intra-industry production will have grown in sectors where the firms of both countries involved have a relative technological advantage.

There is an additional reason to expect intra-industry production in such a case. This is connected to what (at the start of Chapter 2) was described as the third proposition of the theory of technological accumulation, namely that technology is differentiated between firms and locations. If so, firms may wish to directly establish production in a foreign centre of innovation in order to gain access to a potential source of technological development which is distinctive to firms operating in that location but complementary to their own. Certain aspects of innovation which are specific to the foreign location can then be incorporated by the firm into a broadening of its own path of technological development. In such cases intra-industry production will tend to replace intra-industry trade, as the most innovative firms ensure that they expand their research and production in all the most important locations for technological activity in their sector.

This gives rise to three hypotheses which are tested in this chapter. The first is essentially an alternative to a hypothesis that can be found in the literature on intra-industry trade, in which literature it is sometimes postulated that intra-industry production will be highest in the same industries in which intra-industry trade is highest. The alternative advanced here is that there need be no such necessary connection between the sectoral pattern of intra-industry trade and production, once it is allowed that they may be substitutes as well as complements (as argued above). Following on from this, the second hypothesis is that intra-industry production is most likely to substitute for intra-industry trade the greater is the role of technology creation in any sector. Third, it is expected that where the firms of two countries are strongly engaged in technological competition, the international production they each locate in the other country will be particularly attracted towards sectors in which domestic firms have their own technological advantage.

Intra-industry production occurs where firms from two different countries but from the same industry each establish production facilities in the other country. Where international technological competition between MNCs whose home countries are centres of innovation prevails, then intra-industry production will steadily increase. To give an illustration, suppose that four of the major centres for innovation in the pharmaceuticals industry lie in West Germany, Switzerland, the UK and the USA. If the pattern of innovation in each is distinctive, then MNCs from each of these four countries will be increasingly attracted to establishing production in the other three. If this happens then intra-industry production between these four countries in the pharmaceuticals sector will gradually increase. While due to their technological advantage in the sector US MNCs will be heavily involved in international economic activity in the pharmaceuticals throughout the rest of the world (see Chapter 6), their international production will be particularly attracted to the UK, West Germany or

Switzerland. Under these conditions of gaining access to new sources of technological activity the growth of intra-industry production may also be associated with the decentralisation of R&D activities within the MNCs concerned, even though strategic control of new technological developments may remain at corporate headquarters.[1]

This is one aspect of the possibility formulated by the present author elsewhere (Cantwell, 1987; and Dunning and Cantwell, 1989) and suggested by the analysis of Chapters 3 and 4, that production agglomerates in certain sites through a process of cumulative causation. The most dynamic locations continually attract new production facilities and upgrade their existing activity, while in the least successful locations plants close and production stagnates for lack of new investment. The strongest firms in an industry are all drawn to the main international centres of innovation, and in the process they help to further strengthen the local technological capacity of such centres. At the same time production is drawn away or relegated to assembly types of activity in other sites, accompanied by a dwindling in the amount of local research carried out in the sector in question. A model which illustrates this will be developed in Chapter 8.

Where it exists, this tendency towards virtuous circles in some locations at the expense of vicious circles in others can be explained in part by the requirements of technological accumulation in the MNCs of a global industry. The technology accumulated by firms producing in each location is to some extent specific to that location. It is dependent upon the type of innovations previously established there, the skills developed by indigenous workers, the characteristics of the local education system, and the nature of linkages that exist between firms in the industry and with local suppliers and customers. In the most successful locations for any industry technology accumulates rapidly and domestically operating firms gain the capacity to become MNCs, as they begin to expand into international markets. In order to transplant the technology that they have accumulated at home to production in a foreign location, it is in general advantageous for them to exercise control over that foreign production (Pavitt, 1987). Not only have they acquired a great deal of experience in the use of that technology which cannot be easily transferred to foreign firms accustomed to working with a different type of technology, but they are also better placed in their ability to further develop their own particular form of innovation. However, although each firm follows its own distinctive path of technology creation there are likely to be overlaps between them, and hence scope for technological joint ventures in which each party contributes something different, and may make a different use of the results.

In order to establish production in the new location, further local innovation is required so that products and processes can be adjusted to their new operating environment. The technology used must be adapted to local conditions, including a distinctive set of suppliers, customers and final product markets.[2] The appropriate form of production is also affected by the state of infrastructure, and the skills and training of the local labour force, as well as by local customs, tradition, and national legal framework. The use of technology in a new location cannot be dissociated from the creation of complementary technologies in the same location to make its implementation effective, even if this must be done in cooperation with local firms. By adapting their technology to local conditions firms ease the way for cooperation with local companies. The process of adaptation involves drawing on local types of technology and integrating them with those of the MNC, taking technological development in a direction which both expands the horizons of the investing firm and widens local capability. By the same token, if firms that are innovating successfully in one location wish to also draw on the complementary innovations that might be more easily developed in an alternative location, then they must enter into foreign production.

Moreover, in recent years there has been a growing convergence of formerly separate branches of innovation, giving rise to what have been termed technological systems or galaxies. This has not only increasingly brought MNCs into competition with one another, but it has meant that what were once alternative streams of technological accumulation in different MNCs have become progressively more complementary to one another. This has enhanced the tendency to create poles of growth and poles of stagnation amongst the locations of production in international industries. To continue their existing path of technological accumulation in such a way that they remain competitive with their major rivals in an industry the strongest MNCs must produce in the domestic locations of these competitors, which have provided the latter with an environment conducive to their own particular type of innovation. Countries with the strongest traditions for innovation in an industry will not only be the homes of their own MNCs, but they will also be the hosts to the MNCs of other such countries who wish to gain access to a potential source of innovative activity complementary to their own. The trend will be for the strongest MNCs in each sector to locate both research and production facilities in all the major sites of innovative activity in their industry. The new demands of technology creation within firms in an era of global competition require them to establish international productive networks, at a minimum across those countries in which technological innovation is favoured in their industry. Once some MNCs begin to do this, expanding their technological base and thereby the complexity of their production and the extent of their gains from economies of scope, others cannot afford to be left behind for too long.

This in turn suggests that the phenomenon of 'intra-industry production' is likely to continue to spread. The view that intra-industry production may expand as a result of a process of technological competition between MNCs in an international industry gives rise to the three hypotheses outlined above, which are tested once again using evidence on the six major industrialised countries and their firms. It is not intended to provide a general explanation of all intra-industry production, but simply to help clarify an important element in the recent internationalisation of industry, reflecting the way in which technological accumulation has become internationalised alongside capital accumulation. This particular motive for a growth in intra-industry production is compatible with other explanations which have emphasised oligopolistic interaction in which major rivals invest in one another's domestic markets (Graham, 1985; Sanna Randaccio, 1980).

7.2 Some Evidence on Intra-industry Production and Trade in the Industrialised Countries

This section examines the evidence on intra-industry production and trade (two-way trade flows in the same sector) in Europe, the USA and Japan using the sources of data described above. There is, however, a central problem with empirical work in this area. That is, as had been noted already, data on international production are rarely available at a detailed industrial level. For this reason, as in previous chapters, it proved necessary to work with a cross-section of just 12 manufacturing industries. This is unfortunate, since in general not only does the statistical measure of intra-industry trade or production fall as the degree of industrial disaggregation increases, but the sectoral variation becomes wider and more complex. This means that the results reported below must be treated with a certain amount of caution.

The measure of intra-industry trade (IIT) used here is that suggested by Grubel and Lloyd (1975). It is an index calculated for any country across i industries (where in the calculations below $i = 12$), and it depends upon exports from each industry i (X_i) and imports (M_i). The index is given by:

$$IIT_i = [(X_i + M_i) - | X_i - M_i |]/(X_i + M_i)$$

Likewise, the index of intra-industry production (IIP) depends upon the outward international production of the firms of that country (OP_i) and the inward international production of affiliates located in the country but owned by foreign MNCs (IP_i). This index is given by:

$$IIP_i = [(OP_i + IP_i) - | OP_i - IP_i |]/(OP_i + IP_i)$$

Both indices vary between zero and unity. At zero there is no intra-industry trade or production, while at one all international trade or production is defined as being of an intra-industry kind.

In work on intra-industry trade the IIT index is sometimes adjusted (as proposed, for example, by Aquino, 1978) to allow for an overall imbalance, whether in the form of a surplus or deficit, in total trade in manufactures. The greater is the extent of the overall imbalance the lower will be the maximum possible average value of IIT or IIP as defined above. However, as Greenaway and Milner (1986a) point out, an overall imbalance in international trade (or production) may well be the 'normal' state of affairs for many countries. The danger of the Aquino method is that the very factors which explain intra-industry production and trade, and the cross-industry variation in the IIP and IIT indices, may be removed by the adjustment procedure. There is even less reason to believe in a natural balance between total outward and inward production that there is in the case of exports and imports. A country which is a substantial net outward or net inward investor will indeed have little intra-industry production precisely because of this, and this should be reflected in the measure employed. For the purposes of the discussion here, the unadjusted Grubel and Lloyd index of IIP and IIT is therefore the most appropriate one.

With reference to the introductory comments of Section 7.1, the index of intra-industry production may be high under two quite different sets of circumstances. In the case of an attractive location it may be high because both outward and inward international production are strong, while an unattractive location may score a high value because both outward and inward international production are weak. This implies that where there is an increasing polarisation in an international industry between attractive and unattractive locations, there will be a general rise in the index of intra-industry production. In countries which are homes to the most innovative MNCs in any given industry there will be a tendency towards greater inward investment by foreign-owned MNCs, while in countries with little or no indigenous research capacity the inward investments of foreign MNCs may be subject to rationalisation.

Although the IIP index rises across all industries as the patterns of outward and inward international production become more similar, it tends to do so more systematically in sectors in which the home country is an attractive location relative to those in which it is unattractive. When both outward and inward international production are strong the index is more likely to be high as the denominator ($OP_i + IP_i$) will be high and compared to this there is less scope for departures in the relative magnitudes of OP_i and IP_i ($|OP_i - IP_i|$). When both outward and inward international production are weak then the IIP index may assume a low value if either is close to zero. It is therefore not surprising that Dunning and Norman (1986) report that their measure of intra-industry direct investment (a proxy for IIP) for the USA in 1980 is correlated with the sectoral shares of both outward and inward investment. In other words, IIP_i is more likely to be high where both outward and inward international production are buoyant than where they are both weak.

The index of intra-industry production in 1982 for the four largest European countries, and for the USA and Japan, is shown

Table 7.1   The index of intra-industry production, 1982

	USA	West Germany	UK	Italy	France	Japan
1. Food Products	0.53	0.34	0.32	0.22	0.67	0.28
2. Chemicals	0.90	0.70	0.81	0.83	0.79	0.77
3. Metals	0.81	0.73	0.66	0.49	0.64	0.55
4. Mechanical Engineering	0.63	0.74	0.97	0.17	0.34	0.46
5. Electrical Equipment	0.49	0.97	0.70	0.52	0.80	0.11
6. Motor Vehicles	0.34	0.94	0.91	0.92	0.94	0.16
7. Other Transport Equipment	0.89	0.80	0.37	0.06	0.98	0.91
8. Textiles	0.67	0.86	0.26	0.14	0.38	0.32
9. Rubber Products	0.37	0.55	0.67	0.61	0.53	0.10
10. Non-metallic Mineral Products	0.86	0.77	0.29	0.12	0.70	0.28
11. Coal and Petroleum Products	0.64	0.05	0.43	0.60	0.94	0.49
12. Other Manufacturing	0.74	0.62	0.66	0.03	0.16	0.01
Total (national balance)	0.69	0.79	0.60	0.65	0.88	0.38
Mean (of 12-industry index)	0.66	0.67	0.59	0.39	0.66	0.28

Source: As for table 5.1.

in Table 7.1. The average value of the index is highest for France and West Germany, lower for the UK, and lower still for Italy. The mean value of the IIP_i index across the distribution of industries is a measure of the overall significance of intra-industry production for the country in question. This mean value of the index over the 12 industries is shown in the final row of Table 7.1. By comparison, when the formula for IIP is applied to total outward and inward international production then the result is a measure of the national balance between the two (as reported in the penultimate line of Table 7.1), which reflects just one possible contributory factor in intra-industry production. As already mentioned, an imbalance between outward and inward investment will tend to lower the IIP index, but even where no such imbalance exists the average value of IIP_i may still be very low. Where total outward international production is exactly equal to total inward international production (the 'national balance' is one) this may be either because of an exact balance between comparatively advantaged and comparatively disadvantaged sectors each of which has little or no intra-industry production, or because of a high degree of intra-industry production within each sector. For countries in which total outward international production is catching up with total inward or vice versa, the national balance will move closer to one but the measure of intra-industry production will only definitely do likewise if outward investment rises fastest in those sectors in which inward investment currently exceeds it by most.

Consider firstly how the national balance has affected the extent of intra-industry production in the six countries. France, West Germany and Italy are all hosts to a value of foreign-owned affiliate production that is greater than the value of international production that is controlled by their own MNCs. In all three countries the outward investment of their own MNCs has been catching up the more established position of inward investment, but this has reached a more developed stage in the case of French and West German firms than it has with Italian. The UK is rather different, as a country with substantial net outward international production. Many large UK MNCs are highly competitive in their industries, and they have been able to expand their international production at a time when the UK has been becoming a less attractive location for production (as outlined in Chapter 5).

The USA has a mean value of its index of intra-industry production which is around the level of France and West Germany. Like the largest UK firms, US MNCs are often well established and have sustained a high level of international production; but unlike the UK, the USA has become a more attractive location for the production of non-US MNCs. Intra-industry production in Japan, on the other hand, is in general at a very much lower level than for any of the European countries. The international production of Japanese MNCs has begun to catch up with that of their major European competitors, but US and European MNCs have for various reasons found it much more difficult to establish production in Japan.

Turning to the sectoral variation in the index, there are three European industries in which intra-industry production is especially significant. These are chemicals, electrical equipment and motor vehicles. Above-average values of the index are recorded for these sectors by all four European countries. These high values are a consequence of both high outward and inward international production in each case, with the exception of motor vehicles in the UK which has lost out as a result of rationalisation at a European level, and of the French chemicals sector which is rather average. This suggests that West Germany, the UK, Italy and France, and their firms are all involved in international oligopolistic competition in these sectors. The USA and Japan also have relatively high intra-industry production in chemicals, and in this case global competition extends beyond Europe. This contrasts with electrical equipment and motor vehicles in which Japan and the USA have low levels of intra-industry production. Although US MNCs have a reasonably strong position in these sectors, and Japanese MNCs have a very strong position, European MNCs have been much less able to penetrate the American and Japanese markets in these cases.

Intra-industry production is on the whole comparatively low in Europe in food products, metals, textiles and non-metallic mineral products. The exceptions here are the UK metals sector, and the West German textile industry. These are both fairly average sectors that do not have especially high or low outward or inward international production, and which have below-average measures of intra-industry production. This suggests the conclusion that global competition is not terribly strong as a locational influence in Europe in any of the four industries mentioned. One qualification to this is that UK MNCs are very strong in the food industry, and UK and French MNCs account for a high proportion of total international production in non-metallic mineral products (essentially building materials), but in each case their home countries are not critical production locations in these industries. At the other extreme, Italian firms are very strong in the domestic production of textiles, but as yet they have little international production. In the first case competition has in a sense been raised to a global level for a country's firms but not for the country itself, while in the second the existence of a wider international industry has become more important for the country than for its firms.

The index of intra-industry trade for 1982 can be found in Table 7.2. As might be expected, overall manufacturing trade is more nearly in balance for countries than is international production. West Germany, Italy, France and Japan were surplus countries, the USA was in deficit, and the UK roughly in balance (though the UK has more normally been a deficit country in the 1980s, so the year 1982 was unusual in this respect). Japan holds a much stronger trade position than any of the European surplus economies. It can be seen, though, that a more balanced position in a country's trade or production need not necessarily imply a higher

Table 7.2   The index of intra-industry trade, 1982

	USA	West Germany	UK	Italy	France	Japan
1. Food Products	0.82	0.68	0.69	0.71	0.86	0.17
2. Chemicals	0.77	0.72	0.81	0.87	0.91	0.46
3. Metals	0.59	0.76	0.99	0.79	0.93	0.41
4. Mechanical Engineering	0.64	0.44	0.75	0.59	0.97	0.28
5. Electrical Equipment	0.91	0.84	0.88	0.95	0.93	0.21
6. Motor Vehicles	0.61	0.35	0.85	0.88	0.86	0.03
7. Other Transport Equipment	0.38	0.89	0.71	0.71	0.71	0.21
8. Textiles	0.48	0.81	0.75	0.47	0.86	0.76
9. Rubber Products	0.93	0.78	0.89	0.61	0.77	0.24
10. Non-metallic Mineral Products	0.80	0.87	0.99	0.43	0.93	0.30
11. Coal and Petroleum Products	0.54	0.48	0.86	0.83	0.60	0.06
12. Other Manufacturing	0.89	0.85	0.80	0.64	0.84	0.58
Total (national balance)	0.07	0.83	1.00	0.86	0.88	0.60
Mean (of 12-industry index)	0.70	0.71	0.83	0.71	0.85	0.31

Source: United Nations, *Yearbook of International Trade Statistics*, 1984.

average value of the intra-industry index (since it is possible to have perfectly balanced trade or production in aggregate with no intra-industry component). Indeed, there is little difference in the average value of the indices of intra-industry trade and production for West Germany, the USA or Japan, but intra-industry trade is on the whole greater than intra-industry production for the UK, Italy and France.

The sectoral variation in the intra-industry trade index is rather more nationally distinctive than for intra-industry production, and it is harder to pick out sectors for which intra-industry trade is clearly important or unimportant at a European level. It appears to be greatest as a rule for chemicals, metals and non-metallic mineral products. In motor vehicles intra-industry trade is also quite high, except for West Germany which holds a strong net export position. Taking both intra-industry trade and production into account, it is in the chemicals and motor vehicles sectors that global competition seems to be most significant for Europe and her firms.

7.3 The Relationship between Intra-industry Trade and Production

There is a growing literature that suggests a close connection between intra-industry trade and production.[3] In the context of an industry characterised by technological competition between MNCs it is possible that the emergence of intra-industry trade, as firms attempt to gain access to the markets of their major rivals, is gradually displaced by intra-industry production as they attempt to gain access to the development of new technology.

If intra-industry production increases as a result of global technological competition, then such a trend may offset an otherwise positive correlation between the indices of intra-industry trade and production. This is because in these circumstances MNCs set up production in the domestic locations of their major rivals in order to extend their capacity to innovate, and not simply to enter the home markets of these rivals more effectively. To the extent that there is an association between the two, the mutual penetration of MNC investments is liable to substitute for intra-industry trade. This substitution effect may counteract the consideration (advanced, for example, by Caves, 1986) that intra-industry trade and production are likely to be high in similar kinds of industries; for example, industries with a high degree of product differentiation.

Whether intra-industry trade and production are substitutes or complements depends upon which of two possible types of intra-industry production prevails. The first type involves horizontally integrated international production, one motive for which is that described above, namely that technology creation may be specific and differentiated in each location. As MNCs come to service a global market rather than a series of independent national markets, such differentiated technological accumulation may become more internationally integrated through the coordination of research and learning. In this case, intra-industry production may replace intra-industry trade. Apart from other instances of oligopolistic interaction between major rivals, a similar outcome may also result where protectionism increases, or the size of national markets is enlarged or extended through the creation of customs unions, or the role of scale economies declines, or transport costs change. In some industries, the growth of intra-industry production in place of trade may therefore be driven by trade policy rather than by the requirements of technological competition.

The other possible type of intra-industry production might be thought of as not genuinely intra-industry. It occurs when different stages in a vertically integrated chain of production are located in different countries, and linked through intra-firm trade. Firms based in different countries are each responsible for coordinating similar such international networks. This second kind of intra-industry production is of necessity accompanied by intra-industry trade, so where it is relatively important the IIP and IIT indices will be correlated with one another. However, it might be argued that the intra-industry character of such trade and production is a statistical illusion, since it only appears at a sufficiently high degree of sectoral aggregation. Once final and intermediate products are distinguished through a more refined disaggregation then it becomes inter-industry trade and production. The debatable inclusion of this kind of trade and production as intra-industry is known in the literature on intra-industry trade as the problem of categorical aggregation (Greenaway and Milner, 1986a).

The relative importance of the two types of intra-industry production can be tested by assessing the extent of correlation between IIP and IIT indices. The first kind of horizontally integrated production may be associated with either a positive or negative correlation between the two, whereas the second kind of vertically integrated production is likely to generate a clear and positive correlation between them.

The degree of correlation between intra-industry trade and production was tested using simple cross-section regression analysis. Using the 12-industry index the following relationship was estimated for each country for 1982:

$$IIT_i = \alpha + \beta IIP_i + \varepsilon_i$$

The results are reported in Table 7.3. A positive correlation between intra-industry trade and production exists for three out of the four largest European countries, but it is only significant in the case of Italy. For France and the USA there is if anything an inverse relationship between the two. In order to assess the overall position of the four European countries a composite index was calculated by adding together the individual country indices. Hence,

Table 7.3   The results of the regression of the intra-industry trade index for 1982 on the intra-industry production index for 1982

Country	$\hat{\alpha}$	$\hat{\beta}$	t_α	t_β	R^2
West Germany	0.753	0.036	2.66 *	0.12	0.001
UK	0.957	0.292	10.14**	0.32	0.010
Italy	0.648	0.290	9.40**	3.16**	0.500
France	1.141	−0.219	10.02**	−1.66	0.217
Europe	3.002	0.169	5.74**	1.09	0.106
USA	0.904	−0.183	4.62**	−0.93	0.079
Japan	0.420	0.013	3.01 *	0.12	0.001

*   Denotes coefficient significantly different from zero at the 5% level.
**   Denotes coefficient significantly different from zero at the 1% level.
Number of observations = 12.

where all four countries have a low value of intra-industry trade or production then so does the composite index for Europe, and vice versa. As might be expected, the combined indices of intra-industry trade and production for Europe are positively, but not significantly, correlated with one another. This finding is in line with that of Dunning and Norman (1986) who report no significant correlation between intra-industry direct investment and IIT indices at a more disaggregated 56-industry level for the USA in 1980.

It appears, at least at the existing level of disaggregation, that there is only a slight and not a strong association between intra-industry trade and production. This may mean that it is the first type of intra-industry production that is more important, that which allows for the possibility that intra-industry production displaces rather than is accompanied by intra-industry trade. Elements of substitutability seem to be as significant as complementarity between IIP and IIT. This leads into the question of the circumstances under which intra-industry production is likely to be high or low relative to intra-industry trade. There may be sectors in which MNCs are less heavily reliant upon trade between the major areas of competition and market growth, so that the geographical spread of production is more dispersed. In the context of technological competition, it is likely that international production will grow relative to trade where an industry is an important source of innovation for other sectors (for example, a capital goods industry or a sector in which core technological skills are developed). Firms are then required to maintain close links with their suppliers and customers in the country in which they operate, in order to make their contribution to a continuing chain of innovation.

The form of such vertical inter-firm linkages may vary between countries, especially where technology creation is location-specific. Local component suppliers may be able to provide new innovative equipment and other inputs that the firm has not been accustomed to working with elsewhere, and which require the two to maintain regular contact and liaison, and perhaps even causes the firm to enter into a wider range of subcontracting arrangements than it had originally intended. For their part, local customers may use the equipment they are provided with in a production process which is different from that of the customers that the firm supplies in other countries, and so the equipment is subject to a different set of tests. The feedback that the firm gains from such customers is critical to the further development of its product, and by helping the customer to test various possible modifications it accumulates new technological experience. Moreover, it is not simply a case of having to establish production facilities in order to be able to respond to the needs of local suppliers and customers. It is precisely in these cases that the MNC stands to gain most from complementary foreign innovation, and through its experience of local production can engage in reverse technology transfer.

If this argument is correct then where technological competition is a factor in explaining intra-industry production, it will be especially prone to give rise to intra-industry production rather than trade where the sector in question is important as a source of technology creation for the firms of other sectors. To test this view an index of the significance of manufacturing sectors as net creators of innovations was compiled, and is shown in Table 7.4. The index is derived from the work of Scherer (1984) on inter-industry technology flows. It is based on data on R&D expenditure and patented innovations in the USA in 1974 and 1976–7 respectively. The total value of the technology created by each industry was measured by the value of its R&D expenditure, and this was then allocated between the groups of patented innovations for which firms in the sector were responsible. This was done in a US Federal

Table 7.4   The index of innovation creation relative to usage based on US company-financed R&D, 1974, and patents, 1976–1977

1. Food Products	0.8503
2. Chemicals	2.3589
3. Metals	1.2079
4. Mechanical Engineering	5.8073
5. Electrical Equipment	3.7548
6. Motor.Vehicles	4.9270
7. Other Transportation Equipment	2.7895
8. Textiles	0.7234
9. Rubber Products	0.8932
10. Non-metallic Mineral Products	1.1422
11. Coal and Petroleum Products	0.7660
12. Other Manufacturing	2.1199
Total manufacturing	2.2556

Source: Scherer (1984).

Trade Commission survey of companies, in which firms were asked to report the relative magnitudes of R&D spending needed to bring each type of patented product or process to fruition. Scientists and engineers then estimated the usage likely to be made of products embodying the resulting technology by the firms of different sectors. By this means the value of technology created by each sector was then apportioned according to its use by firms in the same or other sectors. This allowed Scherer to build up a technology input–output matrix.

The index reported in Table 7.4 represents the total value of the technology created by each sector divided by the total value of technology used by that sector. It can be seen by the figure for all sectors that manufacturing industry is a substantial net creator of innovations; Scherer's overall matrix incorporated the net use of technology by enterprises in the extractive and services sectors. The relationship between this index of net innovation creation (NETINN) and the ratio between the composite indices of European intra-industry production (EIIP) and intra-industry trade (EIIT) was estimated. The following regression result was obtained:

$$EIIP_i/EIIT_i = \quad 0.683** \quad + \quad 0.105** \ NETINN_i$$
$$\qquad\qquad (7.32) \qquad\quad (3.17)$$

$$R^2 = 0.501 \quad ** = \text{significant at the 1\% level on a}$$
$$\text{two tailed } t \text{ test}$$

This result provides support for the hypothesis that the more important a sector is as a source of technology creation for other sectors, the greater will be intra-industry production relative to intra-industry trade. Where firms are required to establish close linkages with others in a chain designed to encourage innovation, then each of the leading MNCs in the industry will attempt to disperse its production more widely across the main sites of innovative activity. In order to coordinate and strategically guide this kind of locationally dispersed technological and productive activity, global organisation will be more common in such industries. This is consistent with the finding of Dunning (1980) that intra-industry production has been rising fastest in just such global industries, in which international investment of a 'rationalised' kind plays an especially significant role.

7.4   Technological Competition between MNCs from West Germany and the USA

In order to advance the argument further it is necessary either to extend the degree of sectoral disaggregation (which the lack of availability of data prevents), or to examine in rather greater detail the pattern of international production in a bilateral case. To follow the latter course data are needed not only on the industrial distribution of the total international production of a country, but also on the destination of outward production for each industry, and on the country of origin of inward production. In this way the industrial characteristics of production in each country by the firms of the other can be assessed. Since this type of data on outward and inward international production is available only for the USA, comparisons can only be carried out for intra-industry production running between the USA and selected European countries.

In Chapter 6 it was shown that the total international economic activity of West German and US firms is significantly correlated with their own respective patterns of technological advantage. West German firms appeared to be more highly dependent upon technological advantage than the firms of any other European country. As there is reason to believe, therefore, that West German firms are particularly involved in technological competition, the discussion that follows is based on West German and American evidence. What is of interest here is the geographical distribution of the international production of the firms of these countries, one component (with exports) of their total international economic activity. Do they tend to locate such production in other international centres of innovation in each industry? If they do, then there should be some correlation between the industrial distribution of the production of West German-owned foreign affiliates in the USA and the industrial pattern of technological advantage of indigenous US firms. A similar relationship should apply between the production of US MNCs located in West Germany and the technological advantages and innovative activity of West German firms.

An index of the importance of the international production of West German firms in the USA relative to their total international production was therefore constructed. Likewise, another index was calculated as a means of measuring the extent of the attractiveness of West Germany for the international production of US MNCs in each sector. The values of these indices for 1982 are shown in Table 7.5. The construction of these indices was based on the assumption that acquiring access to a source of fresh innovation is only one of a number of possible locational attractions that may draw MNC investment into a country, but that it is an especially important influence for West German and US MNCs, given their relatively high degree of involvement in technological competition. Accordingly, the first index measures the extent to which West German MNCs are attracted to the USA relative to all non-US MNCs, and the second index is concerned with the relative attractiveness of West Germany as an industrial location for US MNCs compared to all non-West German MNCs.

The index of the attraction of West German firms' international production to the USA (GIPUS) depends upon the production of West German-owned foreign affiliates in the USA in industry i ($GFAUS_i$), the production of all foreign affiliates in the USA ($TFAUS_i$), the total international production of West German firms in the world ($GFAW_i$), and the total international production of all non-US MNCs in the world in the industry in question (total international production in the world minus that of US MNCs,

Table 7.5 The index of the relative attractiveness of the USA for the international production of West German firms, and of West Germany for the international production of US firms

	GIPUS	USIPG
1. Food Products	2.349	0.238
2. Chemicals	0.921	0.668
3. Metals	0.755	0.526
4. Mechanical Engineering	0.551	0.785
5. Electrical Equipment	0.416	0.814
6. Motor Vehicles	0.883	1.674
7. Other Transportation Equipment	1.805	1.231
8. Textiles	0.466	1.828
9. Rubber Products	2.121	1.200
10. Non-metallic Mineral Products	0.536	0.632
11. Coal and Petroleum Products	13.734	0.486
12. Other Manufacturing	0.353	0.631

Source: US Department of Commerce, *Foreign Direct Investment in the United States, 1980*, October 1983, and *US Direct Investment Abroad: 1982 Benchmark Survey Data*, December 1985; Deutsche Bundesbank, *Supplement to the Monthly Report on the Balance of Payments*, Reports Series 3, no. 3, March 1985; and as for table 5.1.

$TFAW_i - USFAW_i$). The index of the relative attraction of US firms' international production to West Germany (USIPG) depends upon US-owned foreign affiliate production in West Germany in industry i ($USFAG_i$), total foreign affiliate production in West Germany ($TFAG_i$), the total production of US-owned foreign affiliates throughout the world ($USFAW_i$), and the total value of international production in the world less that of West German firms ($TFAW_i - GFAW_i$). Using these terms, the equations for each index are as follows:

$$GIPUS_i = (GFAUS_i/TFAUS_i)/[GFAW_i/(TFAW_i - USFAW_i)]$$

$$USIPG_i = (USFAG_i/TFAG_i)/[USFAW_i/(TFAW_i - GFAW_i)]$$

In other words, the first index shows the share of German MNCs in international production located in the USA, relative to their share of the total international production of non-US firms in the world as a whole. The second index reflects the share of US-owned foreign affiliates in international production in West Germany, relative to the US share of the total international production of non-West German firms in all locations. The indices shown in Table 7.5 were calculated for 1982. They show that West German MNCs were particularly attracted to the USA in the coal and petroleum products, the rubber and plastic products, the food products and other transportation equipment sectors; while the international production of US MNCs was especially attracted to West Germany in the cases of motor vehicles, textiles, rubber and plastic products and other transportation equipment.

The hypothesis tested here is that West German MNCs are especially attracted to producing in the USA, and US MNCs particularly favour West German locations, in those sectors in which the firms of the other country have established distinctive innovative advantages of their own. The test relies upon the by now familiar cross-industry measure of the technological advantages of West German and US firms. To recap, it is an index of revealed technological advantage (RTA) in 1972–82, which has been constructed using data on patenting activity in the USA in which US and West German MNCs figure prominently. The relevant values of the index are shown in Table 6.1. The test was conducted through a regression of the GIPUS index on the technological advantage of US firms, and of the USIPG index on the technological advantage of indigenous West German firms, the expectation being a positive relationship in each case.

Allowance was also made for the role of 'push' as well as 'pull' factors causing firms to have a higher propensity to locate production in the home countries of their major competitors. While Chapter 6 has shown that firms may have greater international production in the world as a whole in sectors in which they are technologically advantaged, they may be particularly likely to do so in countries in which local firms constitute a major source of technological competition. This may make GIPUS dependent upon the RTA of West German firms, and USIPG dependent upon the RTA of US firms (an additional influence, since GRTA and USRTA are not positively correlated with one another). In fact, in the USIPG equation the coefficient on USRTA though positive was not significant, and its inclusion can be rejected on an F-test. This implies that in sectors in which West German firms are not comparatively technologically advantaged, West Germany is not such a generalised source of research and scientific expertise that despite this advantaged US firms are particularly keen to be represented there. The final results were as follows:

$$\text{GIPUS}_i = -52.12^{**} + 44.03^{**} \text{ USRTA}_i + 9.49^* \text{ GRTA}_i$$
$$(-4.90) \qquad (6.05) \qquad\qquad\quad (2.49)$$

$R^2 = 0.853$ $**$ = significant at the 1% level on a two-tailed t test
$*$ = significant at the 5% level on a two-tailed t test

$$\text{USIPG}_i = -0.88^* + 1.85^{**} \text{ GRTA}_i$$
$$(-2.00) \quad (4.12)$$

$R^2 = 0.630$ $**$ = significant at the 1% level on a two-tailed t test
$*$ = significant at the 5% level on a two-tailed t test

This supports the hypothesis that West German and American MNCs are positively attracted to locations which are important sites of innovative activity in the industry in which they are involved. Moreover, advantaged West German firms are to a lesser extent attracted to the USA in general as a broad centre of innovative activity. This is a very interesting result which serves to further undermine the assumption that is still widespread in much of the literature, that the role of the foreign subsidiaries of MNCs in innovation is to act simply as agents for the implementation and diffusion of technology transferred to them by parent companies. If this were the case then those MNCs that most depend upon technological advantage (such as those of West Germany and the USA) might have been expected to have biased the location of their international production away from other centres of innovation, where they would meet the strongest competitive challenge. If instead, the foreign affiliates of MNCs involved in global techno-logical competition have a role in the creation as well as in the diffusion of technology, then this helps to explain their location in the home territories of major rivals.

7.5   Conclusions

The standard literature on technology diffusion (and on international innovation diffusion and the MNC) is still frequently set
out, whether implicitly or explicitly, around the notion of a
sequence that runs from technology creation to transfer (to an
affiliate or a licensee) to diffusion (to a wider group of firms). The
applicability of this view to innovation in modern international
industries has recenly been challenged (for example, by Dunning
and Cantwell, 1989). An alternative perspective can be found in
certain work (such as that of Lall et al., 1983) on Third World
multinationals, in which the concept of reverse technology transfer
has been developed in the context of their (admittedly small)
investments in industrialised countries. It is reported that some
Third World MNCs have located production in Europe and the
USA as a means of acquiring technology that can be used by their
parent companies.

A more general alternative arises when MNCs are regarded as
being reliant upon a process of technological accumulation.
MNCs can be thought of as representing alternative international
networks of such accumulation, that then enter into technological
competition with one another. Each MNC follows its own specific
path of gradual technological development, which can be
broadened by producing and hence gaining new innovative experience in different locations. European and industrialised country
MNCs have their own form of 'reverse technology transfer' as part
of a two-way international exchange. Their objective is to remain
competitive through the cumulative extension of a network of
international production and innovation. Within such a network
the creation and diffusion of technology are interdependent, and it
may be misleading to think of them in terms of an individual
sequence.

MNCs produce in centres of innovation in their industries in
order to gain access to new sources of technology creation, as well
as to more effectively implement their own variety of technology in
a different environment. The evidence presented in this chapter
suggests that this is an important element in the recent development of international technological competition. In particular, it
has been a major factor in the growth of intra-industry production, which is a vital aspect in sustaining such competition. For

Europe, the USA and their firms technological competition and the international location of innovative activity has been a significant influence in the rise of intra-industry production.

NOTES

1 Evidence that this is so is beginning to appear. For example, the proportion of R&D expenditures undertaken by US-owned foreign manufacturing affiliates in the R&D spending of the total corporate groups rose from 6.6% in 1966 to 8.8% in 1982 (Pearce, 1986).
2 In the case of the semiconductor industry, Flaherty (1983) considers evidence that local production may lead to supporting R&D facilities.
3 See, for example, Greenaway and Tharakan (eds, 1986), Greenaway and Milner (1986a and b), Casson et al. (1986), Helleiner (1981) and Helpman and Krugman (1985) on the intra-firm trade of MNCs as a component of intra-industry trade, and Erdilek (ed., 1985) and Dunning and Norman (1986) on intra-industry production with references to intra-industry trade.

8

A Classical Model of the Impact of International Trade and Production on National Industrial Growth

8.1 Introduction

Earlier chapters have shown how the rapid growth of international trade and production in the post-war period has created a system of international industrial competition, and how this has been associated with an extension in the geographical network of technological accumulation on the part of the firms involved. This chapter presents a model of the impact of the cumulative international growth of MNC production and technology on the competitive position of countries in terms of their shares of production and trade in each industry. The model is constructed on the basis of the findings of the book thus far, and it extends them in certain directions. Unlike the model of Chapter 3, the predictions derived from it are not tested here, but they point the way for future research continuing the lines of argument advanced in earlier chapters.

The conditions which are assumed to underpin the working of the model are therefore closely related to the propositions tested above. The industrial distribution of innovation amongst the firms of each country is supposed to follow an established pattern of accumulated comparative advantage in technological change (Chapter 2). This then regulates the host country competitive impact of MNC growth (Chapters 3 and 4), the composition of

MNC growth itself amongst the firms of each country (Chapters 5 and 6), and helps to create intra-industry production in areas of mutual strength (Chapter 7). The model develops the argument that this pattern of cumulative growth works to the benefit of strong locations (or centres of innovation) in an international industry, at the expense of the weaker.

The foundations are provided by a basic model in which the growth of output is related to the local rate of innovation and hence productivity growth. Once international trade and hence competition between alternative sources of supply is allowed for, the growth of output may rise where innovation is relatively higher and fall where it is lower. International competition then intensifies where international trade increases (only discussed briefly in earlier chapters, but linked with MNC growth in the post-war period), and international production and the international development of technology increases (as described in Chapters 5 to 7). This intensification of international competition is argued to enchance a pattern of cumulative causation in which centres of innovation enjoy a virtuous circle, while certain other locations are as part of the same process locked into a vicious circle of declining growth and diminishing local research.

As mentioned at the start of Chapter 7, the types of activity that MNCs organise in each location may be different. Centres of innovative activity prove attractive to the production and research of foreign MNCs, while there are other locations that attract only assembly types of production and not research, and as a result an international process of cumulative causation can be set in motion or reinforced (Cantwell, 1987). Once the most successful MNCs in a given international industry concentrate research in particular locations then such favoured sites are more likely to experience a virtuous circle of cumulative success, in which their world share of innovation and production steadily rises. In an established sector this is likely to be achieved at the expense of a vicious circle of cumulative decline in research activity in certain other less attractive locations.

This model presented here formalises the connection between the international technological and capital accumulation of firms, and the diverse effects on national industrial growth in different locations. To do so, it uses a classical economic model of the effects of international trade on growth, which is broadened out to consider international production as well as trade. The model is 'classical' in that the focus is on the growth path of each industry over time, which is sustained by means of the reinvestment of the surplus or profits that are generated by production. It is also 'classical' in that, as in the models of Smith and Ricardo, international trade influences profitability and domestic economic growth over time (Myint, 1977; Thweatt, 1976; Walsh, 1979; and Cantwell, 1986c). In each industry it is supposed that the capital accumulation generated by the reinvestment of profits is linked to innovation or technological accumulation, which raises the productivity of labour over time.

The analysis proceeds by means of a multi-sectoral model. For each country, technological accumulation is comparatively high in some sectors, but comparatively low in the others. In an industry in which productivity growth is relatively high the country gains a competitive advantage in international trade, and its strong net export position, by raising the demand for domestically produced output, increases the growth of output and productivity still further. International production strongly reinforces this effect, by way of its differential impact on technological accumulation in different countries, to the advantage of established centres of innovative activity.

8.2   The Basic Model

The foundation of the analysis is a classical growth model into which the effects of international trade and then production are introduced, under conditions of differential rates of innovation across countries. The basic model is regulated by a growth cycle, in which the economy in question alternates between periods of rapid growth or prosperity, and periods of slower growth or stagnation. The idea of a growth cycle was originally formulated by Goodwin (1967), and has been further developed by the same author more recently (Goodwin and Punzo, 1987).

Suppose that the world economy consists of m countries denoted by $i = 1, 2, \ldots, m$, and k industries denoted by $j = 1, \ldots, k$. Taking Q_{ij} as the value of net output (after capital depreciation) in country i in industry j; N_{ij} as the number of workers employed in each sector; A_{ij} as labour productivity; W_{ij} as the value of wages paid out; Π_{ij} as the value of profits; K_{ij} as the value of the capital stock accumulated out of past profits; x_{ij} as the share of wages in net output; k^* as the constant capital–output ratio; w_{ij} as the wage rate; r_{ij} as the rate of profit on capital invested; L_{ij} as the total sector-specific labour force which grows at a constant rate g_{ij}; and v_{ij} as the proportion of the labour force employed; then the basic system is described by the following nine identities or definitions:

$$Q_{ij} = A_{ij} N_{ij} \qquad (8.1)$$
$$Q_{ij} = W_{ij} + \Pi_{ij} \qquad (8.2)$$
$$K_{ij} = k^* Q_{ij} \qquad (8.3)$$
$$W_{ij} = x_{ij} Q_{ij} \qquad (8.4)$$
$$\Pi_{ij} = (1 - x_{ij})Q_{ij} \qquad (8.5)$$
$$w_{ij} = W_{ij}/N_{ij} \qquad (8.6)$$
$$r_{ij} = \Pi_{ij}/K_{ij} = (1 - x_{ij})/k^* \qquad (8.7)$$
$$L_{ij} = f_{ij}^* e^{gt} \qquad \text{where } f_{ij}^* \text{ is constant} \qquad (8.8)$$
$$v_{ij} = N_{ij}/L_{ij} \qquad (8.9)$$

It is supposed that labour is not homogeneous across sectors, but requires training and the acquisition of certain skills specific to an industry, and therefore wage rates and the growth rate of the labour force vary between sectors. However, it is likely that the growth of the labour force (g_{ij}) will be higher where technological accumulation and output growth is higher, and thus the increase in the demand for labour is rising most rapidly, which could be incorporated explicitly in a more sophisticated treatment. It is also assumed that wages are paid at the end of each production period out of net output, leaving a surplus which is paid as profits and is available for capital accumulation. To simplify matters further, it is supposed that all profits are reinvested, so if investment in each sector is given by I_{ij} then an investment equation may be written as follows:

$$\Pi_{ij} = I_{ij} = dK_{ij}/dt \qquad (8.10)$$

The assumption underlying equation (8.3), which shows a constant capital – output ratio, is that capital accumulation and output growth tend to run alongside one another at more or less the same rate. This implicitly supposes that industries which enjoy a faster rate of process innovation (or technological accumulation) also tend to have a higher rate of product quality improvements which ensures that the demand for their output grows faster (perhaps related to a gradual substitution away from the products of other sectors). The growth equation of the system can be derived from equations (8.3),(8.7), and (8.10), using dots above letters to indicate proportional rates of growth (for example, $\dot{Q} = (1/Q)(dQ/dt)$):

$$\dot{Q}_{ij} = \dot{K}_{ij} = r_{ij} = (1 - x_{ij})/k^* \qquad (8.11)$$

Now output growth and capital accumulation depend upon the rate of technological accumulation. It is innovation which propels the system forward over time. It works by raising the share of profits or surplus available for accumulation, lowering the share of wages denoted by x_{ij}. Faster productivity growth generates faster output growth, though this in turn feeds through into a more rapid rate of rise of wages. However, it is supposed that the increase in wages runs behind output and productivity growth, so long as there is sufficient unemployment. When labour shortages are encountered then the share of wages may rise again, and this acts as a constraint on growth. To describe this further, suppose that the rate of technological accumulation is given by $\dot{T}_{ij}$, and that it enters the specification of a technical progress function as follows:

$$\dot{A}_{ij} = u_{ij}{}^* + \dot{T}_{ij}\,\dot{Q}_{ij} \qquad \text{where } u_{ij}{}^* \text{ is constant,}$$
$$0 \leq \dot{T}_{ij} < 1 \qquad (8.12)$$

This function is based on an assumption that higher output growth is associated with higher productivity growth in any industry, but that for any given rate of output growth productivity will rise at a faster rate the greater is technological accumulation. There is also an autonomous element of productivity growth, unrelated to output growth or technical progress, and this is denoted by $u_{ij}{}^*$. Finally, the system is closed by relating the change in the share of wages over time to the degree of 'capacity utilisation' or to the proportion of the labour force employed. Where a higher proportion of the labour force is employed then the share of wages rises more rapidly (or falls more slowly), as the growth of labour demand becomes more constrained by the availability of labour supply. The labour market equation is as follows:

$$\dot{x}_{ij} = b_{ij}{}^* \, v_{ij} - h_{ij}{}^* \qquad \text{where } b_{ij}{}^* \text{ and } h_{ij}{}^*$$
$$\text{are constants} \qquad (8.13)$$

Now the growth in the share of the labour force employed is easily obtained. Using equations (8.1), (8.8), (8.9), (8.11), and (8.12) it is clear that:

$$\dot{v}_{ij} = \dot{N}_{ij} - \dot{L}_{ij} = \dot{Q}_{ij} - \dot{A}_{ij} - g_{ij}$$
$$= (1 - \dot{T}_{ij})\,(1 - x_{ij})/k^* - (u_{ij}{}^* + g_{ij})$$

The system works as a high value of the proportion of the labour force employed feeds through into a positive growth in the share of wages in output, which in turn by slowing the rate of economic growth pulls down the proportion of the labour force employed, maintaining a cyclical process in motion. The position of this cycle, in terms of the feasible values of the share of wages and the proportion of the labour force employed, can be shown by considering the conditions under which the share of wages and the proportion of the labour force employed are in (temporary) equilibrium. This is where their rate of change over time is zero.

$$\text{Where } \dot{x}_{ij} = 0 \qquad v_{ij} = h_{ij}{}^*/b_{ij}{}^* \qquad\qquad (8.14)$$
$$\text{Where } \dot{v}_{ij} = 0 \qquad x_{ij} = 1 - k^*(u_{ij}{}^* + g_{ij})/(1 - \dot{T}_{ij}) \quad (8.15)$$

While the combination of these two conditions for a point of equilibrium is feasible it is dynamically unstable, and the economy will not tend to gravitate towards it over time. However, these equations help to define the shape and position of a 'limit cycle' which describes the motion of the model economy over time, and which is illustrated in Figure 8.1. Although the central equilibrium point is unstable, the system is globally stable along the path of the cycle.

The position of this cycle of activity is lower, the greater is the rate of technological accumulation or the value of T_{ij}, and so a faster rate of innovation is associated with a lower average share of wages, and hence a higher average growth rate. Figure 8.1 shows that the value of x_{ij} specified in equation (8.15) can be thought of as the average share of wages in the economy over time, and where T_{ij} is higher this will be lower. From equation (8.11) it is clear that

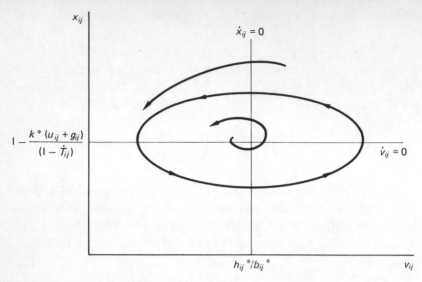

Figure 8.1   The cyclical growth path of an innovating sector of an economy, based on the relationship between the share of wages in output and the proportion of the labour force employed.

the average rate of growth of output and the rate of capital accumulation is greater where the average share of wages is lower. The faster overall rate of technological and capital accumulation both combine to generate higher productivity growth, as shown in equation (8.12). Moreover, to the extent that a distinction between 'high growth' and 'low growth' sectors becomes established historically, a high growth sector is likely to call forth a faster growth of suitably trained labour, g_{ij}, and this further lowers the average share of wages in equation (8.15) and hence raises the growth path of the sector in question.

Since labour is treated as sector-specific, the wage rate varies across sectors. The rate of growth of the wage rate in each industry can be found by combining equations (8.1), (8.4), (8.6), (8.11), (8.12), and (8.13). The resulting expression is as follows:

$$\dot{w}_{ij} = \dot{x}_{ij} + \dot{A}_{ij} = b_{ij}{}^{*} v_{ij} + \dot{T}_{ij} (1 - x_{ij})/k^{*}$$
$$+ (u_{ij}{}^{*} - h_{ij}) \qquad (8.16)$$

It is likely that in innovative high growth sectors (which have relatively high values of T_{ij} and low values of x_{ij}) that the growth of the wage rate tends to be higher, unless the conditions of labour supply happen to be very much easier so that v_{ij} remains relatively low.

8.3   The Impact of International Trade with Differential Rates of Innovation

Now suppose that the model that has been outlined so far applies to the world economy and that the countries concerned participate in trade with one another. The existence of international trade means that production in different locations is brought into competition. However, if autonomous productivity growth (not related to technological or capital accumulation) and the rate of growth of labour supply are equivalent in any industry across countries, then the growth of production in any location depends upon the rate of innovation. With reference to equation (8.15), if $u_{ij}{}^* = u_j{}^*$ and $g_{ij} = g_j$, then the level of x_{ij} and hence the rate of growth and all the other variables of the system depend upon the rate of technological accumulation.

The model as described so far is satisfactory on one condition. This is that the rate of technological accumulation in any industry must be the same in every location, or $\dot{T}_{ij} = \dot{T}_j$. In this case the average rate of growth or capital accumulation in any sector is equal across all countries. To make matters simpler still, also assume that the other key parameters of the system are identical in each country, $b_{ij}{}^* = b_j{}^*$ and $h_{ij}{}^* = h_j{}^*$, so that the extents of fluctuations in growth are the same too. Even though countries may be trading with one another, in the event of common rates of technological progress it is not necessary to deal explicitly with international trade or relative prices, as they do not affect the rate of growth. This depends upon an assumption that has so far been implicit, namely that the ratio of the money wage rate to the real level of labour productivity is set by the rate of productivity growth, which is the same where the rate of innovation is the same.

The underlying intuition here is that a higher level of productivity in real terms tends to be associated with a proportionately higher level of real wages and, where prices are the same, with a proportionately higher level of money wages. So in the most advanced countries where the level of technological capability is greater the real level of labour productivity is higher, but at the same time output, wages and the wage rate are higher in the same proportion. Providing that a higher level of the wage rate is exactly balanced by a higher level of productivity, then countries are equally competitive in a sector, and they have a common or uniform price level. Taking P_{ij} as the price of output in country i in industry j; R_{ij} as the mark-up over nominal wage costs; and a_{ij} as the real level of labour productivity or $a_{ij} = A_{ij}/P_{ij}$, then the price equation may be written:

$$P_{ij} = (1 + R_{ij}) \, (w_{ij}/a_{ij}) \tag{8.17}$$

Now from equations (8.1), (8.4), and (8.6) it is clear that the mark-up over wage costs is given by:

$$R_{ij} = (1 - x_{ij})/x_{ij}$$

In other words, with a fixed ratio of the money wage rate to real productivity, the price level depends upon the share of wages in output, which it has been shown depends upon the rate of technological accumulation. So a uniform rate of technical change, at the same time as setting a common share of wages in output, also establishes a uniform price level. Where the level of productivity is higher the real wage and thus the money wage also rises proportionately higher. The possibility of a 'wage–price spiral' in which money wages rise ahead of productivity, in turn raising prices and making domestically produced output less competitive, is not considered. It is instead assumed that in the modern world economy the major influence on price competitiveness is the rate of technological change, particularly in manufacturing sectors.

The other assumption that it is necessary to make in this connection is that the price elasticity of demand in each sector is equal to unity. If this holds, then changing prices in line with the change in the share of wages called forth by the given rate of technological accumulation exercises no separate effect on the value of output.

It is not, however, supposed that the ratio of the money wage rate to real labour productivity is the same in every sector. Industries which sustain a faster rate of technological progress and therefore productivity growth have a lower wage to productivity ratio, and a lower level of prices. The idea underlying this is that although a higher level of productivity tends to be associated with higher wages, that productivity growth tends to run ahead of wage growth, gradually pulling wages up behind it. If so, productivity growth exerts a downward pressure on prices through this route until wages catch up. With international competition production sited in locations in which productivity growth is fastest (due to the most rapid rate of innovation) gains a competitive advantage. Assuming that prices are set in accordance with the general ratio of wages to productivity elsewhere in the world, if this ratio is lower in the most innovative countries then their effective mark-up over costs (R_{ij}) increases. Technological progress thereby has an indirect effect on the share of wages (x_{ij}) through the competitive advantage that it creates by lowering the wage–productivity ratio compared to the rest of the world, in addition to the direct effect that it has on the share of wages which has been discussed in Section 8.2.

So long as the rate of technological accumulation in an industry is uniform across countries, then under the assumptions stated above no location has a competitive advantage in international trade, and the share of wages and the rate of growth in capital and output in the sector is everywhere the same. Moreover, since the comparative rate of innovation across sectors is the same in every location, there is no comparative advantage in international trade as the sectoral structure of relative prices is identical in each country. Comparative levels of productivity may vary, but where this happens comparative wage rates vary in the same fashion. It is comparative rates of productivity growth which are equivalent, so long as the same sectoral rate of innovative advance prevails in every location.

So, beginning with a world economy in which in any international industry technological accumulation proceeds at the same rate across countries, then there are no locational competitive advantages, and trade is in balance in each sector in every country. In the terminology of Chapter 7, there is no inter-industry trade, though there may be intra-industry trade (which is not considered in this model, as it would require further disaggregation to bring it into the story).

This ceases to hold once differential rates of technological progress are allowed for, as in some locations productivity growth begins to run faster, but prices do not fall in the same proportion as they are held back by slower productivity growth elsewhere in the world. The effect of this is to increase the nominal mark-up over wage costs in the most innovative locations. This increases the share of profits or lowers the share of wages in output, and raises the rate of capital accumulation and economic growth above what is achieved in other countries. Note that, even though the share of wages falls, because output growth increases as well, wages may still grow faster in a 'progressive' economy of this kind.

To illustrate this process it is easiest to make the simplifying assumption that with the introduction of differential rates of technological change the international pattern of consumption in nominal terms continues to grow in each location at a rate that reflects the domestic share of wages (which are spent on consumption) and profits (which are invested) rather than their share in the world as a whole. This is once again associated with the assumption of a unitary price elasticity of demand, such that the nominal level of domestic consumption is not affected by charging a price related to the world ratio of wages to productivity rather than the internal ratio. It is the location of production and output, not consumption, which is affected by differential productivity growth. Under these conditions, where productivity growth is relatively high it is output but not consumption which rises faster (as a higher value of output is matched by a lower share of wages), opening up a trade surplus in the industry equal to the difference between output and consumption.

Now let consumption in country i in industry j be denoted by C_{ij}; the prevailing world price is given by P_j; use P_{ij} as the notional domestic price that would apply if the world average productivity growth and thus the ratio of wages to productivity were equal to those that exist in country i (so, for example, if technological accumulation is above that in the rest of the world then this notional price lies below the actual price, $P_{ij} < P_j$); let x_{ij} be the share of wages in net output, such that in the case of high innovation $x_{ij} < x_j$ due to an adjustment in the mark-up over costs where unit costs are lower; and let X_{ij} be the value of exports, and M_{ij} the value of imports. Three identities can then be written which are additional to the basic system described by equations (8.1) to (8.12) above:

$$C_{ij} = (P_{ij}/P_j)Q_{ij} \tag{8.18}$$
$$W_{ij} = x_{ij}Q_{ij} = x_j\,(P_{ij}/P_j)Q_{ij} \tag{8.19}$$
$$X_{ij} - M_{ij} = Q_{ij} - C_{ij} \tag{8.20}$$

From equations (8.18) and (8.19) it follows that $P_{ij}/P_j = x_{ij}/x_j$. This result should be intuitively clear from the explanation above. Because productivity growth varies across countries in any international industry, so does the wage to productivity ratio, but since prices are maintained through competition at a common international level adjustment takes place through the mark-up over costs and hence in the relative share of wages and profits, rather than through the formation of a different national price. For this reason the share of wages is raised or lowered in the same proportion that prices would have changed, had they adjusted to the new domestic ratio of wages to productivity. The expression for the share of wages relative to the international average can be substituted into equation (8.18), which in turn substitutes into equation (8.20) to give:

$$X_{ij} - M_{ij} = Q_{ij} - (x_{ij}/x_j)Q_{ij} = [1 - (x_{ij}/x_j)]Q_{ij} \quad (8.21)$$

This is subject to the condition that trade in an international industry must balance, in the sense that total surpluses must equal total deficits, or $\Sigma_i(X_{ij} - M_{ij}) = 0$. This implies that:

$$\Sigma_i x_{ij} = \Sigma_i x_j = mx_j \quad (8.22)$$

A competitive advantage in international trade obtained through a relatively fast rate of technological accumulation generates a faster rate of domestic growth and a trade surplus. Now it is possible that the trade surplus itself becomes responsible for a yet higher rate of growth, and hence a process of cumulative causation. The opening up of trade surplus represents a faster growth in the demand for domestically produced output. It is clear from equation (8.12) that the higher growth of output compounds the effect of speedier technological change to generate a yet faster rate of productivity growth. The equation may be rewritten:

$$\dot{A}_{ij} = u_j^* + \dot{T}_{ij}(1 - x_{ij})/k^* > u_j^* + \dot{T}_j(1 - x_j)/k^* = \dot{A}_j$$

Starting from a position in which technological change and the rate of growth are equivalent across countries, if the rate of technological accumulation increases in country i then although initially $x_{ij} = x_j$, the faster rate of innovation feeds through into faster productivity growth. This, however, it has now been shown, leads to a lowering of x_{ij} below x_j, which creates still faster productivity growth and in turn a further reduction in the share of wages, x_{ij}. The increase in growth achieved by an improved competitive position in world markets feeds upon itself. Higher productivity growth produces a greater mark-up over costs, a lower share of wages, and an increased growth rate through the opening up of a trade surplus, but this generates yet higher productivity growth and so the process goes on. Left to itself, growth continues to increase in the most innovative and successful locations, but continues to fall in the least innovative. Alongside this, the size of trade surpluses expands where technological accumulation is at its strongest, while trade deficits widen where it is at its weakest.

In a world economy of this kind, the proportional rate of change of the share of wages depends upon the current level of the share of wages. Where they fall below the average industry-wide share the associated increase in output growth raises productivity growth, which pulls down the share of wages further below the industry average. The opposite happens where the share of wages rises above average in a less innovative location. To formalise the dependence of the rate of change of the share of wages on their level, suppose that the growth of the wages share is related not only to the proportion of the labour force employed, but also to the size of the trade imbalance relative to the size of output. In particular, suppose that the share of wages in output falls faster (or rises slower) in inverse proportion to the magnitude of the trade surplus relative to output, as follows:

$$\dot{x}_{ij} = b_{ij}^{*}\, v_{ij} - h_{ij}^{*} + z_{ij}^{*}\, [1 - (x/x_{ij})]$$
$$h_{ij}^{*} > z_{ij}^{*} > 0 \tag{8.23}$$
$$\text{Where } \dot{x}_{ij} = 0 \qquad v_{ij} = [(h_{ij}^{*} - z_{ij}^{*})$$
$$+ z_{ij}^{*}\, (x/x_{ij})]/b_{ij}^{*} \tag{8.24}$$

The condition that $h_{ij}^{*} > z_{ij}^{*}$ is required as it is only where the share of the labour force employed is positive that it is feasible to have a positive and stable share of wages in output. However, in this new state cumulative departures away from the average share of wages in output which prevails in other countries are possible.

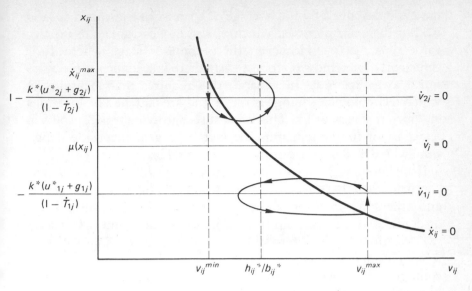

Figure 8.2 The growth path of production in separate locations with different rates of techological accumulation

It is only where technological accumulation in country i takes place at the international 'average' rate ($\dot{T}_{ij} = \dot{T}_j$) and so the share of wages is also average ($x_{ij} = x_j$) that the system collapses into that described in Section 8.2. Equations (8.13) and (8.14) now appear as special cases of equations (8.23) and (8.24). As technological progress rises above the international average productivity growth increases, and improved international competitiveness is reflected in a higher mark-up over unit costs, or a lower share of wages in output. The trade surplus that accompanies higher growth feeds through into a further fall in the share of wages, and as x_{ij} continues to drop below x_j, it is necessary for a progressively larger share of the labour force to be employed to prevent x_{ij} from slipping back further. This is illustrated in Figure 8.2.

Once the rate of technological accumulation and hence the share of wages and the growth rate depart from the international norm, then the course followed by the economy depends upon the extent of that departure. Assuming for a moment that $u_{ij}{}^* = u_j{}^*$ and $g_{ij}{}^* = g_j$, then if the difference between $\dot{T}_{ij}$ and $\dot{T}_j$ is relatively small the national sector in question may simply follow a distorted cycle. In the case of a high rate of innovation the cycle lies below the cycle described by the average of the rest of the world (x_j and v_j), but there are points at which x_{ij} still moves above the average value of x_j, $\mu\,(x_j)$. However, there comes a stage where $\dot{T}_{ij}$ is sufficiently far removed from $\dot{T}_j$ that there is a tendency for x_{ij} to spiral away from x_j. In general, this requires that locations in which technological change is very rapid are matched by locations in which it is very weak. This is because a growing trade surplus in one part of the system must be met by a growing trade deficit somewhere else, as set out in equation (8.22).

If technological innovation is sufficiently differentiated across countries then there are some in which x_{ij} steadily falls and v_{ij} rises, and others in which the reverse is the case. However, the process of cumulative success and decline is not without limit. Consider two locations in the international industry j, where innovation is relatively high in country 1 but relatively low in country 2. In country 1 x_{1j} moves downwards and capital accumulation rises, but eventually further increases in growth are blocked when the share of the labour force employed (v_{1j}) hits its maximum value, v_{ij}^{max}. In the limit the maximum value of v_{ij} is unity when the whole of the labour force is employed, but in practice the capacity constraint is likely to be reached before this. At this stage the large trade surplus cannot continue to promote the high rate of growth, and so the final term in the expression for $\dot{x}_{ij}$ drops out again, and equation (8.23) lapses back to equation (8.13). The consequence is that x_{1j} rises back to the $\dot{v}_{1j} = 0$ line in Figure 8.2, at which point the pressure on capacity is removed. The process then begins again, with the economy following a cycle that periodically runs up against supply constraints, with growth sustained at an above-average rate.

At the other extreme country 2 sees the share of the sectoral labour force employed fall to its minimal point, $v_{2j} = v_{ij}^{min}$. This minimum is greater than zero if local producers are to remain in business. Whether they do or not depends upon the maximum value of x_{2j}. When x_{2j}^{max} approaches or exceeds one then profits are zero or negative, growth is negative, and local firms are liable to be forced out of business. If instead it is simply a point of slow growth (rather like the 'stationary state' of the classical economists) then employment does not fall any further and the trade deficit ceases to exercise downward pressure on growth and hence the share of the labour force employed. Here again, equation (8.23) reverts to equation (8.13) until the very low value of v_{2j} pulls the share of wages back down to the $\dot{v}_{2j} = 0$ line. There is then a cycle with relatively high values of x_{ij} and relatively low values of v_{ij}, as shown in Figure 8.2.

Under these circumstances an international industry becomes divided between three kinds of location for productive activity. There are the poles of high growth and low growth, and a third group of countries clustered in the middle between them. This form of separation into three distinct groups may only clearly begin to take effect where there is a substantial differentiation in rates of technological progress. It is also more likely when the growth of the sector-specific labour force, g_{ij}, or autonomous productivity growth, u_{ij}^*, are higher where innovation is greater, and vice versa. Where there are only minor differences in rates of innovation then growth paths may be pulled away from that illustrated in Figure 8.1, but they do not tend to spiral up or down towards the maximum or minimum limiting points. Even where some countries do witness such a dramatic effect, there remain a possibly quite large group of countries that stay in the middle or 'average' range of growth.

Comparing the growth of the wage rate between countries with different rates of innovation requires an expression that is rather more complicated than equation (8.16). Drawing on equations (8.1), (8.4), (8.6), (8.11), (8.12), and (8.23), this is:

$$\dot{w}_{ij} = b_{ij}^* \, v_{ij}^* + z_{ij}^* \, [1 - (x_j/x_{ij})] + \dot{T}_{ij} \, (1 - x_{ij})/k^* \\ + (u_{ij}^* - h_{ij}^*) \tag{8.25}$$

Equation (8.25) can be used to compare the rate of growth of the wage rate in highly innovative and less innovative locations for production. Compare an average economy in industry j with technological accumulation $\dot{T}_j$ and share of wages in net output x_j with a faster growth country, with $\dot{T}_{ij} > \dot{T}_j$ and $x_{ij} < x_j$. To simplify matters assume that the autonomous element of productivity growth, the constant term in the proportional rate of change in the share of wages equation, and the rate of growth of the sector-specific labour force are all equivalent across countries; that is, $u_{ij}^* = u_j^*$, $h_{ij}^* = h_j^*$, and $g_{ij} = g_j$. Disregarding for a moment the effect of differences in the proportion of the labour force employed ($v_{ij} > v_j$), the rate of growth of the wage rate is faster in the innovative economy where:

$$[\dot{T}_{ij}(1 - x_{ij}) - \dot{T}_j(1 - x_j)]/k^* > z_{ij}^*(x_j - x_{ij})/x_{ij}$$

Now consider an 'average' point at which $x_{ij} = 1 - k^* (u_j^* + g_j)/(1 - \dot{T}_{ij})$ and $x_j = 1 - k^* (u_j^* + g_j)/(1 - \dot{T}_j)$. Substituting these values into the expression above, it can be shown that on average the rate of growth of the wage rate is faster in the high growth economy when:

$$(1 - \dot{T}_{ij}) > (u_j^* + g_j)k^*/(1 - z_{ij}^* k^*) \qquad (8.26)$$

Equation (8.26) shows that an above-average rate of technological accumulation in a particular location tends to increase the growth of the wage rate where autonomous productivity growth (u_j^*), the growth of the labour force (g_j), the capital – output ratio (k^*), and the sensitivity of profitability and hence capital accumulation to the trade surplus (z_{ij}^*) are all lower. In addition, as is clear from Figure 8.2, where technological progress is greater the proportion of the labour force employed (v_{ij}) tends to be higher, and this also tends to raise the growth of the wage rate depicted in equation (8.25). So a progressive or relatively innovative economy may well combine an improvement in the incomes and living standards of its workers with a comparatively low and possibly falling share of wages in output.

Capital accumulation and output growth are definitely driven higher by faster technological accumulation, but pulled down lower by a weak rate of innovation. In the limit, when technological change is strongly differentiated between countries, some economies may enter a phase of cumulative success at the expense of others who experience cumulative failure. At one extreme the national industry gravitates towards a capacity constraint, and is restricted by labour shortages, while at the other extreme the country finds itself in the classical 'stationary state' following a stagnationist slow-growth or no-growth path with widespread surplus labour.

8.4   The Impact of International Production with Differential Rates of Innovation

The model described in Section 8.3 is a general one which demonstrates the possibility of cumulative causation so long as countries which are trading with one another innovate at different rates in the same industry. So far nothing has been said about the organisation of firms within the industry. If firms produce only within their own national boundaries then it may be that the model shows what is merely a long-run process in which it takes time to penetrate foreign markets and gain the allegiance of foreign consumers. The process is further constrained if transport costs are high and significant tariff and other trade barriers are in operation.

Once international production is brought into consideration, and especially when the costs of transport and communications between countries are falling, the process becomes much more immediate, and it is likely to lead to the creation of a global industry. As production becomes increasingly internationally mobile the differences between locations are accentuated, and with this firms face an increasing incentive to take strategic decisions on where to produce at a world level. In order to sustain a global base for technological accumulation, all MNCs desire to be represented by expanded research and production facilities in the international centres of innovation in their industries. The rate of technological accumulation rises in favoured locations, which makes them still more attractive to MNCs. Production can be switched from a less innovative to a more innovative location by the direct decisions of MNCs, but these decisions in their turn support or undermine research and innovation in each location.

With a network of international production firms can geographically divide research-intensive and assembly types of production. By concentrating research-intensive production in the main centres of innovation in their industry, firms gain direct access to the main sites of technological development. They also bring new research facilities and the distinctive technological experience of their own company to add to and interact with the existing scientific and research expertise of the host country. Inward international production of this kind raises the value of $\dot{T}_{ij}$ both directly, and indirectly through its impact on increased technological competition which causes indigenous firms to extend their own research activity. As has been argued in earlier chapters, inward international production is most beneficial in areas in which the host country and its firms are traditionally strongest.

Elsewhere production may be either withdrawn or increasingly decoupled from local research, as a consequence of which $\dot{T}_{ij}$ falls. Components generated by research-intensive processes are imported to an ever greater extent. While foreign firms essentially rely on their research facilities abroad, domestic firms find that their local research efforts become less viable. By establishing assembly activities locally, foreign firms are able to more effectively penetrate the host country market, and their foreign research is increased as their world market share rises and provides a high volume of global sales from which to fund it. Confronted with declining market shares and competitors with much greater and broader international technological capacity, indigenous firms may be forced to cut back on research or at the very least to narrow their field of specialisation.

The outward international production of a country's own firms may play a similar role to inward international production, especially where firms are in the early stages of internationalisation. By locating production in other centres of innovation abroad and some assembly activities in other sites, already dynamic companies gain a wider global basis from which to nourish technological accumulation at home. The level of domestic research expands and becomes more complex as it is increasingly integrated with complementary foreign technology. As MNCs, firms increase their capacity to devise broadly based and diversified technological systems through their innovation at home and abroad. This particularly applies where firms are first beginning to move from a strong domestic export position towards the creation of international networks, like the manufacturing companies of Japan.

However, where the firms of a country are established MNCs whose international production developed historically, their home base may since have been run down. Indeed, if the companies were MNCs from the outset as was the case with a number of British companies (for example, in the food sector, in which a great deal of productive activity had always been located outside the UK), then there need be no particular reason for their long-standing outward international production to benefit research at home. In such cases, which it has been shown in Chapter 5 and 6 are more common amongst US and especially UK firms than they are amongst their German and Japanese rivals, a global company system still tends to support and promote research in the main international centres of innovation, but this may not necessarily encompass the home country.

In any case, by raising technological accumulation ($\dot{T}_{ij}$) in locations which are already recognised leaders in applied science, research and development, while lowering the rate of technological accumulation in disfavoured sites, the spread of international production pushes more economies to the extreme positions or growth poles of Figure 8.2. The process of cumulative causation outlined above is strongly reinforced by the emergence of a global industry, and as this happens the advantages of MNCs are further increased *vis-à-vis* purely national firms due to their greater scope of international activity, which in turn ensures that a still higher share of world production is absorbed within the international networks of MNCs, or in smaller firms which collaborate with them. The process of cumulative causation is linked to the consolidation of the global character of the industry. As has been demonstrated in the previous section, where innovative activities are unevenly geographically dispersed, countries in which technological accumulation is greater will benefit from higher productivity growth, more rapid capital accumulation and output growth, and a higher share of the labour force consistently employed.

The picture of the role of MNCs just painted needs qualifying to the extent that in any industry there are some countries who do not have any significant indigenous research capability, or have lost such capacity. Industrialised countries in this position may compete among themselves to try and attract inward international production, as a means of averting further decline in a sector in a time of increased international competition, rather than exacerbating such decline. Even with only assembly types of production, the local presence of MNC network-linked effects may raise technological accumulation ($\dot{T}_{ij}$) through the development of local skills and incremental process and quality improvement. However, this does not lead to the restoration of fundamental research facilities, and local innovation becomes more dependent upon a wider international network which is under foreign control, in contrast to the stronger interdependence that comes about between the main centres of research in the industry.

In general, for the major industrialised countries, the internationalisation of production compounds patterns of cumulative causation in the relationship between national industrial growth and international trade. Only where a vicious circle of relative decline has reached the point where local research of a fundamental kind is no longer viable may the expansion of international production act as a brake on, rather than as a spur to, this process.

8.5   Conclusions and some Policy Implications

The model advanced in this chapter gives rise to two interesting conclusions, both of which are at variance with much of the standard literature on international trade and production. The first conclusion is that the trade patterns of countries, and their gains and losses from international trade, are determined not so much by their absolute levels of technological capability or productivity (as in the traditional Ricardian measure of absolute and comparative advantage), but by their relative rates of innovation or productivity growth. The second conclusion is that trade imbalances between countries must be expected as a normal and not an abnormal state of affairs, particularly as international production increases.

The central feature of the model is that technological accumulation is depicted as the essential underlying source of productivity growth, and this in turn is viewed as the key to international success on the part of countries, industries and firms. It is not the level of productivity that is crucial in competitive success, so much as its rate of growth. This helps to explain why in the post-war period, across a wide range of industries, Europe, Japan and then the newly industrialising countries were able to sustain rapid rates of export growth despite having a level of labour productivity that was below that of the USA. In the sectors in which their activities were most dynamic, their strong rates of innovation meant that productivity rose ahead of wage costs, opening up a position of absolute advantage in world markets. This allowed them to maintain a faster rate of capital accumulation and hence growth of output in their favoured industries. It also explains why industrialised countries whose innovativeness is weakened in a variety of sectors are prone to protectionist lobbies.

The other major characteristic of the model is that, while it is recognised that in an international industry the surpluses of some countries must be matched by the deficits of others (equation (8.22)), there is no reason why in a particular country the surpluses of some sectors must be matched by the deficits of others. All countries have a mix of successful and unsuccessful industries, but they do not automatically balance out in the way that is assumed in conventional (comparative advantage) trade models. It is not the rate of technological accumulation in an industry relative to its rate in other sectors in the same country that matters, but the rate of technological accumulation relative to the general rate in the same industry in the rest of the world. This is increasingly so as industries are internationalised, and production and research is drawn towards major centres of innovation. Countries that are centres for research activity across a broad range of industries tend to run persistent trade surpluses as a result.

In conventional trade models imbalances are speedily eliminated through exchange rate adjustments (or, if exchange rates are fixed, by domestic income or price adjustments). A supposition underlying the formulation of the model above is that exchange rate movements or changes in the terms of trade follow in the wake of trade imbalances, but with a lag in the same way that it was supposed that wages follow productivity growth. This leaves room for a sustained improvement in competitiveness so long as productivity growth is continuous. Of course, once changes in exchange rates or terms of trade are allowed for explicitly they complicate the working of the model, and represent a constraint on the ability of countries to have either a very high or a very low proportion of sectors in surplus. Countries which have a large number of industries in which they have a relatively high rate of innovation by international standards tend to experience a systematic appreciation of their currencies over long periods, while less innovative countries witness persistent trade deficits and long-term currency depreciation. Just as high productivity growth gradually pulls the wage rate up behind it at the industry level, so at the broader national level the overall trade surplus that it generates gradually pulls the value of the domestic currency up behind it (although with much greater fluctuations along the way, once financial markets are taken into account). In this view exchange rate trends are ultimately explained over long periods by the competitive position of countries rather than vice versa.

The most immediate policy implication of the two main conclusions of the model is that countries should concentrate their efforts on assisting the expansion of sectors in which they have their greatest innovative potential, and consequently can achieve a dynamic absolute advantage in trade through steadily rising industrial productivity. Previous chapters have shown that the choice of such sectors is governed by the existing and past technological experience of local firms. This policy suggestion is at odds with the conventional recommendation that countries gain by specialisation in industries in which they currently have a static comparative advantage in terms of existing productivity levels.

The rejection of the traditional trade policy view by the belief that countries should specialise in sectors in which comparative productivity growth is highest, rather than in those in which comparative productivity levels are greatest, is shared by Pasinetti (1981) and Ros (1986), although their reasoning is somewhat different. Pasinetti and Ros both examine the path followed by an economy in a world with differential rates of innovation, *if* trade takes place in accordance with static comparative advantage. Pasinetti then shows that if innovation is heavily concentrated in export sectors (in the terminology of Chapter 2, the degree of technological specialisation is high) then domestic productivity growth leads to a steady decline in the terms of trade, while Ros demonstrates that if innovation is potentially greater in import sectors the growth of output and consumption will be lower with trade than under autarky. In the analysis here it has been shown that, when the pattern of trade is determined by productivity growth rather than levels, trade exercises direct effects on national industrial growth, which either enhance or retard domestic growth rates. It is then in the interests of countries to concentrate on improving the conditions of output growth and trade expansion in those sectors in which they have a comparative advantage in innovation (and hence productivity growth).

The model suggests that such sectors may be capable of especially rapid growth, but that they may therefore be more quickly constrained by shortages of trained labour with suitable skills, leading to the share of wages being pushed back up and impeding growth. It is notable in equation (8.15) that the average growth rate achieved by a national industry rises not only with the speed of technological change, but also with the rate of growth of labour supply (g_{ij}). Governments can assist in increasing the supply of appropriately educated and trained workers, particularly if they act with sufficient foresight. This point is stressed by Freeman and Perez (1989), in relation to the lack of availability of skilled labour in a period of a changing technology paradigm when newly emerging sectors offer the key growth. It is discussed further with reference to the UK's microelectronics sector by Freeman and Soete (1989).

An alternative policy prescription is possible. Countries may attempt to improve their competitive position in world markets, which entails lowering the wage rate to productivity ratio in selected industries, by trying to force down wages by weakening trade union power, relying in part on increased unemployment (a lower level of domestic demand reducing the average value of v_{ij}), and partly on weakening the responsiveness of wage growth to the state of the labour market (the relationship between $\dot{x}_{ij}{}^*$ and v_{ij}). Faced with recurrent current account deficits in a country with relatively few innovative industries, restrictive demand policies may also reduce imports and eliminate the least productive firms. However, if the model is an accurate representation of the world, measures of this kind are unlikely to ensure more than a short-lived competitive advantage. They do little to affect the underlying rate of technological accumulation and productivity growth, and indeed may even affect it adversely as reduced domestic demand has to be offset against any rise in export demand. Innovative foreign competitors quickly renew their challenge through increasing labour productivity ahead of wages.

Although the model is particularly intended to apply to industrialised countries, as an aside a similar point may be equally applicable in the case of developing countries. The newly industrialised countries have proved capable of faster rates of growth than other countries with a comparable level of development (in terms of their levels of labour productivity and real wages), because they have been more innovative and consequently capable of a more rapid rate of productivity growth. The view of many in the industrialised world that their success is based essentially on low wages is consequently very one-sided. Indeed, in a successful process of technological and capital accumulation at a national level, rising real wages play a critical positive role in that they help redirect labour towards the most innovative and the fastest-growing sectors. It may well be that the type of innovation varies between countries at different stages of development, but this is a separate point. At a lower initial stage of development there may be a greater emphasis on the imitation and adaptation of foreign technology, and on organisational rather than scientific innovation through reconstituting the types of production process in use; but in its own way this is just as much innovation, and the accumulation of technology and locally relevant skills.

At any point in time in the model proposed above, it is possible to construct a ranking of countries in which those with higher average levels of labour productivity also have higher average levels of real wages. However, over time the ordering changes, as countries and their industries move up or down the ranking of levels of labour productivity, depending upon the relative rates of technological accumulation and thus rates of productivity growth that they are able to sustain. In their successful sectors countries move up the ranking, achieve an above-average export growth, and attract a larger share of the more beneficial types of international production allied to the establishment and extension of local research facilities. In the course of this process successful economies increase their share of world exports, and enjoy higher rates of profit (and a lower share of wages in output) and faster capital accumulation.

In any international industry, the division between regular trade surplus and regular deficit countries encourages a shift of international production towards the former which are centres of innovative activity. The division itself is further enhanced by the shift of production as the increasing concentration of research supports a wider dispersion in rates of productivity growth between locations. Countries which are innovative leaders across a broad range of industries tend to run overall trade surpluses, while countries with a weak all-round innovative performance suffer from recurrent overall trade deficits. This creates particularly acute protectionist pressures in deficit countries; it is a competitive weakness in international trade which is the principal cause of protectionism. Over longer periods of time the structural imbalance in trade leads to adjustments at both the industry and the country level. In the industry high productivity growth pulls up wage rates, while with a persistent overall surplus the value of the domestic currency tends to appreciate. These two effects combine to increase living standards fastest in those countries whose research and innovation is greatest.

The cumulative expansion and reorganisation of the networks of MNCs and the increasing significance of international trade and production has led to a growing competitive interdependence between production locations. While a long-run process of cumulative causation in the location of activity may always have been gradually at work, it has been strongly reinforced in recent years. This helps to explain the current concerns about 'international competitiveness' which are widespread throughout the industrialised economies. Further research is now needed on the precise nature of the relationship between the competitiveness of firms, the locations in which they choose to concentrate their research-based activity, and the competitiveness of countries in the sector in question.

9

Towards an Evolutionary Theory
of International Production

9.1   An Evolutionary Approach

An evolutionary approach is one which examines the steady evolution of a system or a related group of systems over time. The theory of technological accumulation is an example of an evolutionary theory as it argues that the technology of firms and locations is in a process of constant and cumulative change, and its application to MNC activity involves an evolutionary approach. Earlier chapters of this book have shown how the theory of technological accumulation can be extended to the analysis of MNCs and international production. This chapter examines the relevance of the approach that has been taken here for theories of international production.

Chapter 8 summarised some important theoretical implications of the growing connections between the spread of international production in manufacturing and the international organisation of technological accumulation. It did so by developing a particular model of these connections, in part suggested by the empirical evidence presented in previous chapters. Other models of MNCs and innovation which also incorporate the gradual evolution of production and technology over time are possible, and so the objective of this chapter is to set out some of the more general characteristics of the framework within which they are likely to be developed. By doing so it provides a theoretical conclusion to the book, and acts as an agenda for future research.

Although the ideas of technological accumulation are incompatible with the standard neoclassical theory of international trade and investment (which assumes that technology is everywhere the same and that all firms have access to the same technology), they may be relevant in the context of various other approaches to MNC activity that can be found in the recent literature. Much of this literature has been concerned to answer the question of why MNCs exist, but it is now increasingly turning its attention to another question raised in this book, namely why some MNCs grow faster than other firms in world markets, and why in each industry they grow faster in some locations than others. Thinking of technology and production as steadily accumulating in international networks helps to answer these questions. It also throws some light on existing approaches to analysing MNC behaviour, and on how they might be adapted to consider cumulative evolutionary processes.

International production can be analysed at three levels: macroeconomic (examining broad national and international trends), mesoeconomic (considering the interaction between firms at an industry level), and microeconomic (looking at the international growth of individual firms). It is quite natural that macroeconomic theories of international production have often relied heavily on theories of trade, location and (in the case of FDI) the balance of payments and exchange rate effects; mesoeconomic approaches tend to be derived from industrial economics, game theory and the theory of innovation; while microeconomic thinking is grounded upon the theory of the firm. The approach of this book has been conducted essentially at a mesoeconomic level.

Using this distinction between different levels of analysis, the main theories of international production can be grouped under four headings. These constitute four alternative theoretical frameworks, since approaches within each share certain common theoretical foundations. However, each of them can be further subdivided between particular theories or approaches, and they are not always mutually exclusive. The first two are based on alternative theories of the firm; the market power or Hymer theory of the firm, and the internalisation or Coasian theory of the firm. The third group are macroeconomic developmental approaches, while the fourth (which includes the approach developed here) are based on the analysis of competitive international industries.

There is in addition another still more general framework for the study of international production which draws on a variety of theoretical elements. This has been developed by John Dunning, and it is known as the eclectic paradigm (Dunning, 1977, 1981, 1988a). The eclectic paradigm is not an alternative analytical framework in the same sense, since it incorporates elements from all four types of approach and can be applied equally well at micro or macro levels. It is rather an overall organising paradigm for identifying the elements from each approach which are most relevant in explaining a wide range of various kinds of international production, and in the wide range of different environments in which international production has been established.

Section 9.2 reviews the four major types of approach to international production, together with the eclectic paradigm. Section 9.3 then examines the relationship between different approaches in the context of issues that are raised when the focus is on the dynamic aspects of international production. This leads into a concluding section on how the various approaches have attempted or are attempting to treat the growth of international production, and on how research into the cumulative evolution of MNC activity may proceed in future.

9.2   A Survey of the Major Theories of International Production

The theory of international production dates from 1960 when Hymer, in a doctoral dissertation eventually published in 1976, showed that the orthodox theory of international trade and capital movements did not explain the foreign operations of MNCs (see Dunning, Cantwell and Corley, 1986). In particular, it did not explain two-way flows of FDI between countries, and still less between countries with similar factor proportions. His explanation of why firms move abroad and establish international production was based on a theory of the firm and industrial organisation. Since that time, four major theoretical frameworks for the analysis of MNCs have emerged, and a fifth overall framework which attempts to bring strands from each together. The first two are based on particular theories of the firm, and although their advocates sometimes claim that they are general theories, they are unlike the eclectic paradigm, which is a general all-encompassing framework which need not be tied to any particular theory of the firm or MNC development. The other two frameworks also collect together somewhat different approaches, derived from adaptations of the theory of international trade or economic development and the theory of oligopolistic competition or technological innovation respectively.

The first theoretical framework used to analyse international production is that passed down by Hymer, based on a view of the firm as an agent for market power and collusion. It comes in both non-Marxist and Marxist versions, the latter dating back to Baran and Sweezy (1966). Two of the clearest recent statements of this framework can be found in Newfarmer (ed., 1985) and Cowling and Sugden (1987).

The second is the internalisation approach, based on a Coasian or institutionalist view of the firm as a device for raising efficiency by replacing markets; it has been advanced as a general paradigm by Rugman (1980), though less extravagant claims are made for a similar approach Buckley and Casson (1976 and 1985), Williamson (1975), Teece (1977), Caves (1982), and Casson (1987), amongst others.

Macroeconomic developmental approaches come in various forms, covering the earliest versions of the product cycle model (PCM Mark I) which trace back to Vernon (1966) and Hirsch (1967); the approach of the Japanese economists Kojima (1978) and Ozawa (1982); the investment-development cycle (Dunning, 1982) and stages of development approach (Cantwell and Tolentino, 1987); and – though these are rather different – approaches which deal with the role of financial factors in FDI (Aliber, 1970; Rugman, 1979; and Casson, 1982).

Approaches based on the analysis of competitive interaction in international industries include the technological accumulation approach advanced in earlier chapters. Others that can be considered under this heading are later Mark II versions of the product cycle tradition (Vernon, 1974; Graham, 1975; Flowers, 1976; Knickerbocker, 1973); the internationalisation of capital approach (Jenkins, 1987); and in a development context the work of those whom Jenkins terms neo-fundamentalist Marxists (Warren, 1980).

The final framework is the eclectic paradigm developed by Dunning (1977 and 1988a), which as its name suggests combines elements of all the other four in such a way that it is compatible with various different theoretical approaches. This is the most convenient starting point of such a broad survey.

9.2.1 *The Eclectic Paradigm*

In the eclectic paradigm it is contended that MNCs have competitive or 'ownership' advantages *vis-à-vis* their major rivals, which they utilise in establishing production in sites that are attractive due to their 'location' advantages. According to Dunning, two types of competitive advantage can be distinguished; the first is attributable to the ownership of particular unique intangible assets (such as firm-specific technology), and the second is due to the joint ownership of complementary assets (such as the ability to create new technologies).[1] MNCs retain control over their networks of assets (productive, commercial, financial and so forth) because of the 'internalisation' advantages of doing so. Internalisation advantages arise both from the greater ease with which an integrated firm is able to appropriate a full return on its ownership of distinctive assets such as its own technology, as well as directly from the coordination of the use of complementary assets, subject to the costs of managing a more complex network.[2]

Dunning (1988a) describes the internalisation advantages that result from the coordination of the use of complementary assets as 'the transactional benefits . . . arising from a common governance of a network of these assets, located in different countries' (p. 2). That such benefits can only be enjoyed through coordination within the firm rather than by market coordination is said to be the result of transactional market failure. Three reasons are given for transactional market failure (Dunning, 1988a, b). Firstly, risk and uncertainty may be significant in transactions carried out across national boundaries. Secondly, where there are externalities benefits external to the transactions concerned may not be captured by parties transacting at arms length. Thirdly, there may be economies of scope through the direct coordination of interrelated activities.

More will be said about ownership advantages in due course, since they have been the subject of considerable debate in the literature. However, there are two points that must be clarified at the outset. Firstly, there may appear to be an overlap between those ownership advantages which are due to the joint ownership of complementary assets and those internalisation advantages which derive from the coordinated use of such assets. In fact, the distinction here is rather like the distinction between the advantages of owning particular assets such as patented technology and the internalisation advantages of retaining control over their use in order to ensure that the full return on them is appropriated by the firm that holds proprietary rights.

Nonetheless, while ownership advantages that derive from particular assets can normally (in principle at least) be sold – such as in the licensing of the use of a technology to another firm – there is in general no market for ownership advantages of a more collective kind. Examples of the collective type of ownership advantage are the overall organisational abilities of the firm, the experience and entrepreneurial capabilities of its managers taken together, the reputation and creditworthiness of the firm in international capital markets, its political contacts and its long-term business agreements with other firms. This kind of ownership advantage goes beyond any particular asset or any one individual, and in general cannot be sold outside the firm but is only usable within it.

One such collective ownership advantage is the ability of the firm to generate new technology, which will eventually result in a stream of new ownership advantages of a particular kind. This example will serve to illustrate the distinction between collective ownership advantages at the level of the firm as a whole and the internalisation advantages of the coordinated use of assets which are associated with them. Consider a firm which holds a strong position in a certain branch of the chemicals industry and which uses its innovative potential to expand into the development of a technologically related chemical process. Suppose for the sake of argument that to date scientific and technical effort in the related sector has been concentrated outside the home country of the firm, so that its development work in this area is undertaken primarily in a subsidiary located in some foreign centre of excellence for the process concerned. A specialised R&D unit is established in the foreign country which is linked up with the parent firm's R&D facilities.

Now the ability of the firm to set up production in the foreign country and to initiate a new research programme there is due to its initial ownership advantage, which consists of established technology and an innovative strength in chemicals. The firm then gains internalisation advantages through the coordination of R&D in home and host countries, which extends its original ownership advantage by increasing its capacity to innovate in both related sectors. In other words there is a progressive interaction between ownership advantages (the possession of technology and the ability to innovate) and internalisation advantages (the international coordination of R&D facilities). Ownership and internalisation advantages increase alongside one another in the case of successful international growth.

The second point that requires clarification at this stage is that the concept of ownership advantages is open to two possible theoretical interpretations. As will become clear, the market power theory of the firm perceives ownership advantages principally as anti-competitive devices which act as barriers to entry against other firms. Meanwhile, the competitive international industry approach sees ownership advantages as competitive weapons which sustain a process of competition between rivals. For this reason ownership advantages are 'sometimes called competitive or [sometimes called] monopolistic advantages' (Dunning, 1988a, p. 2).

In conventional neoclassical or industrial organisation analyses of market structure in which competition and monopoly are treated as opposites, to describe advantages as competitive or monopolistic would suggest a contradiction in terms. This is how matters also appear on the whole to the market power school whose work is one variant of the industrial organisation analysis. In their view if the larger firms in an industry have stronger ownership advantages this reduces the number of firms in the sector and increases the extent of collusion amongst those that remain, thereby restricting competition and implying a higher degree of monopoly power.

However, this contrasts with the classical approach to competition which saw it as a process rather than a market structure. In the dynamic view of competition what matters is not the number of firms within an industry but the mobility of resources (within firms, as well as in terms of the entry and exit of new firms) and the balance of forces between firms in an industry. In this context in an oligopolistic industry Jenkins (1987) argues that competition and monopoly coexist. Firms compete through the continual creation of quasi-monopolistic positions, such as the creation of a new technology ahead of the field. It will be necessary to return to this issue in later sections.

9.2.2   Internalisation

There is a particular theory of the firm that derives from the work of Coase and which lays emphasis on the notion of internalisation, and gives it a more restrictive interpretation than is necessary within the eclectic paradigm. This modern theory of the 'internalisation' of markets as it is applied in the case of international production (see Buckley and Casson, 1976) is based on Coase's (1937) criticism of neoclassical economics. The framework of analysis is like the neoclassical theory of trade and investment based on exchange between individuals or groups of individuals, but it introduces the transaction costs of such exchange, which vary in an arm's-length or market relationship as compared with a cooperative relationship. Where the transaction costs of an administered exchange are lower than those of a market exchange, then the market is internalised and the collective efficiency of the group is thereby increased. Apart from the existence of economies of scope across activities, the direct coordination of transactions may reduce the costs associated with information impactedness, opportunism, bounded rationality and uncertainty (for a summary see Caves, 1982).

It is argued that intangible assets such as technology are especially costly to exchange in arm's length transactions. By thinking of the exchange of technology as a transaction that is internalised when the firm has a horizontally integrated network of production, horizontal integration is treated by analogy with vertical integration. Firms that invest abroad in R&D facilities are therefore treated in exactly the same way as firms that invest in a venture to extract natural resources and secure supplies of raw materials; both are internalising markets, and it is simply that they internalise markets for different commodities.

Where markets are internalised through the common ownership and control of the groups that are involved in exchange with one another, the transaction cost approach suggests the appropriate institutional arrangement on which cooperation between the parties is likely to be founded. At one extreme are joint ventures over which the MNC exerts little direct control, or a largely decentralised MNC in which internal markets regulated by transfer prices have replaced external markets, as emphasised by Rugman (1981). Indeed, where the MNC internalises an externality (an external economy or diseconomy) it may create an internal market where no external market existed previously (Casson, 1986). At the other extreme is the organisational structure stressed by Williamson (1975), that of globally integrated multinationals in which control is centralised and hierarchical. This distinction is discussed in Kay (1983).

Of course, there is an overlap of the spectrums spanned by market and administrative coordination, in that where market exchange is characterised by monopolistic or monopsonistic elements, the MNC may exercise control over its contractual partner without resort to 'internalisation'. Strictly speaking, in the transactions cost approach the firm is defined as the direct organiser of non-market transactions. However, the firm or MNC might equally well be defined as the controller and coordinator of an (international) network of production or income-generating assets (Cowling and Sugden, 1987). If so, the firm may exercise control over production which it has subcontracted out, but for which it is the monopsonistic buyer. Transactions are of an external market kind, but production may be controlled and coordinated from a single administrative centre. In this respect, the internalisation framework offers a theory of the choice between different modes of transacting rather than a theory of the (boundaries of) the firm.

To be of use in empirical work this approach needs to be operationalised in a workable model of transaction costs (Casson, 1981 and Buckley, 1983), but the variables thought to be especially significant are the regularity of transactions between the parties, and the complexity of the technology exchanged. Transaction cost analysis may also require adjustment to take account of the distribution of the gains from exchange under different institutional arrangements (as suggested by Sugden, 1983). Thus, the MNC may not favour the most efficient or lowest cost arrangement if its profit share is higher under another. By the internationalisation of production MNCs may weaken the effectiveness of trade union organisation, and increase the share of profits.

Advocates of the internalisation approach have also recognised the possibility that MNCs may increase profits through the restriction of competition in final product markets, and that this may offset the generally superior allocation of resources associated with MNC activity:

> Welfare losses arise where multinationals maximise monopoly profits by restricting the output of (high technology) goods and services . . . where vertical integration is used as a barrier to entry . . . [or] because they provide a more suitable mechanism for exploiting an international monopoly than does a cartel (Buckley, 1985, p. 119).

However, their emphasis is on the organisation of intermediate product markets, and they believe that, even in the presence of monopolistic elements in the final product market, the creation of new internal markets generates sufficient improvements in efficiency that overall cost minimisation remains the overriding motivation of the growth of the firm:

> The internal market . . . in the long run will stimulate both the undertaking of R&D and its effective implementation in production and marketing. Consequently, dynamic welfare improvement is likely to result (Buckley, 1985, p. 119).

9.2.3   *The Market Power Approach*

While in the transaction cost framework the firm is essentially a device for lowering costs and raising efficiency, the alternative theory of the firm used by Hymer sees it as a means by which producers increase the extent of their market power. A definition of market power can be taken from Sanjaya Lall, whose work at the time was identified with this approach:

> Market power . . . may . . . be simply understood as the ability of particular firms, acting singly or in collusion, to dominate their respective markets (and so earn higher profits), to be more secure, or even to be less efficient than in a situation with more effective competition. . . . The concept may, of course, be applied to buyers (monopsonists) as well as sellers (Lall, 1976, p. 1343).

Originally applied to international production by Hymer (1976), this theoretical approach has been used recently by those such as

Savary (1984), Newfarmer (ed., 1985) and Cowling and Sugden (1987).[3]

The main idea is that in the early stages of growth firms steadily increase their share of domestic markets by means of merger as well as capacity extension, and that as industrial concentration (and market power) rises so do profits. However, there comes a point at which it is no longer easy to further increase concentration in the domestic market as few major firms remain, and at this stage profits earned from the high degree of monopoly power at home are invested in foreign operations, leading to a similar process of increased concentration in foreign markets.

The notion that firms everywhere seek out collusive arrangements as the major means by which they keep profits high is reminiscent of Adam Smith, but the market power school have gone further. According to Smith, competition between firms remained a spur to increased investment and technological change, whereas for those who have emphasised the role of market power in MNC activity investment is not so much an independent response to competition as a means of further extending collusive networks. MNCs are believed to invest in foreign operations to reduce competition and increase barriers to entry in their industry, and by increasing the degree of monopoly power they may even (in the longer term) have an adverse effect on the efficiency of foreign plants.

The market power theory of the firm is therefore clearly at odds with the transaction cost minimising or efficiency maximising theory. To the extent that MNCs raise research and productivity in their foreign operations, and improve efficiency through coordinating different types of plant and different types of technology, the effects on profitability in international industries may be ambiguous. Higher internal efficiency within MNCs may increase competition amongst them, making it more difficult for them to divide markets by agreement, and reducing profitability (or at least offsetting the gains due to greater efficiency). If this continued it may act as a disincentive to a further extension of international production, but in this view greater short-run efficiency is to be understood simply as a source of increased market power, which is likely to reduce the extent of investment in greater efficiency in the future.

The market power approach is often associated with the industrial organisation literature, in which it is commonplace to argue that a more concentrated market structure is allied to greater collusion and a higher rate of profit. It should be noted, though, that in Hymer's original version it was a theory of the firm and of the behaviour of the firm rather than a theory of industrial organisation in the modern sense. In Hymer (1976) the firm appears as an active rather than a passive agent. Hymer followed Bain (1956) in viewing the firm as actively raising entry barriers and colluding with other firms in its industry. In the market power theory the primary causal link runs from the conduct of firms to market structure rather than vice versa. MNCs are seen as building up a position of market power at home, and then in their respective international industries. Their movement abroad is hastened by depression in the home market, which may result in part from their own diminishing incentive to invest due to their ever more extensive market power and collusive agreements.

Kindleberger's (1969) interpretation of the Hymer story placed it more firmly in the industrial organisation tradition which revolves around a structure–conduct–performance model. In Kindleberger's restatement the MNC was seen as a function of market structure characterised by monopolistic competition between differentiated products, rather than as an agent involved in oligopolistic interaction with other firms. The more recent writings of Newfarmer (ed., 1985) and Cowling and Sugden (1987) have moved back towards the Hymer stance, in that while their argument is set in an industrial organisation context, they emphasise the (anti-competitive) impact of MNCs on host country market structure.

The use of an industrial organisation context by more recent authors partly reflects a change in the issues and the institutions under study themselves. Hymer's objective had been to investigate why national firms went abroad, rather than to evaluate the operations of existing MNCs. Today the concern is with the way in which international industries are organised. Cowling and Sugden (1987) contend that internationalisation is undertaken not only as a means of increasing the market power of firms in final product markets, but also to raise the share of profits in two ways. Firstly, the greater ability to shift production between alternative locations strengthens the bargaining power of firms in negotiations over wages and conditions of work. Secondly, by 'putting out' work previously done within the firm to a network of dependent subcontractors, both locally and internationally, the position of collectively organised trade unions in large plants is weakened. This is then integrated with the 'monopoly capitalism' argument of Baran and Sweezy (1966) or the stagnationist argument of Steindl (1952): a combination of a rising share of profits and an increasing market power (which reduces the incentive to invest) leads to a slower growth of demand, and secular stagnation eventually at an international level.

Yet Jenkins (1987) notes that the emphasis that Baran and Sweezy and the Marxists that followed them placed upon monopoly, and their downplaying of oligopolistic competition between MNCs, can be traced back to a time when the USA and her firms held a hegemonic position in the world economy. Since then newer MNCs from Europe, Japan and now the Third World have been growing rapidly. The increasing internationalisation of R&D also suggests that MNC growth has helped to sustain technological competition. The use of the market power approach as a *general* theory of international production therefore has a somewhat dated feel about it.

In the internalisation approach the firm grows by displacing markets which operate in a costly and imperfect way, while in the market power theory it is the growth of the firm which is the essential cause of market imperfections and failure. The eclectic paradigm incorporates elements of both these alternative theories of the firm, since it allows that ownership advantages may act as barriers to entry and sources of market power. However, Dunning himself attaches priority to internalisation and supposes that competition is more important than collusion amongst MNCs: 'It is not the orthodox type of monopoly advantages which give the enterprise an edge over its rivals – actual or potential – but the advantages which accrue through internalisation' (Dunning, 1988b). Indeed, the latest terminology in which the eclectic paradigm has been couched (Dunning, 1988a,b), suggesting that the growth of the firm is a function of market failure, owes much to the internalisation approach. While particular ownership advantages and the internalisation advantages of appropriating rents are attributed to structural market failure, collective ownership advantages and the internalisation advantages of coordinating the use of complementary assets are said to be due to transactional market failure.

Despite the priority which Dunning accords to the internalisation over the market power theory of the firm, it would still be wrong to make the eclectic paradigm synonymous with the internalisation approach, as is sometimes done in literature reviews. The more general nature of the eclectic paradigm was mentioned in Section 9.1, and this allows it to give equal weight to theories of macroeconomic locational advantages, and the interaction between the firm and its macroeconomic environment. Thus: 'The theory of foreign owned production stands at the crossroads between a macroeconomic theory of international trade and a microeconomic theory of the firm' (Dunning, 1988b).

9.2.4 *Macroeconomic Developmental Approaches*

Macroeconomic theories of international production are currently at a rather more rudimentary stage of development than are microeconomic theories (for a survey see Gray, 1982). They have tended to attract less attention than theories of the firm since the demise of the product cycle model or PCM, discussed in Chapter 3. However, their origins go back to the same period, the early 1960s. They emerged as a result of criticisms of the traditional theory of international trade, just as Hymer's work represented a criticism of the traditional theory of international capital movements.

The product cycle idea itself was in the first instance a purely microeconomic one which had been familiar for some time in business schools, but which was applied by Vernon (1966) to topical discussions on patterns of international trade and the balance of payments. Vernon's criticism of conventional trade theory also called attention to the rapid rise of international production amongst US firms in Europe at that time. In the neoclassical Hecksher–Ohlin–Samuelson (HOS) theory, trade (or FDI, where capital movements substituted for trade) should be greatest between countries whose proportional factor endowments are most dissimilar. This could not explain the tremendous post-war expansion of trade and investment in manufacturing industry between the USA and Europe. Neither could a theory which assumed that trade automatically balances in accordance with comparative advantage offer any assistance in explaining regular US trade surpluses in the 1950s.

The role of technological factors in addressing these issues had already been clear to some authors in the 1950s. As is well known,

Leontief (1954) referred to the significance of skilled labour in US export industries. Meanwhile, in linking trade imbalances with the monetary side of the balance of payments, Johnson (1958) suggested that the persistence of a dollar shortage in Europe after the Second World War may be explained by the lag with which innovation in Europe followed that in the USA. Then in the early 1960s two important papers on trade theory appeared. Posner (1961) pioneered a 'technology gap' theory of that part of trade based on innovating and learning faster than others; the link with the problems confronted by Leontief and Johnson are evident. At around the same time Linder (1961) argued that the main motor of trade was a similarity of income levels and patterns of demand, suggesting that trade flows are greatest between countries with similar factor endowments.

The PCM attempted to combine elements from both the Posner and Linder theories of international trade, and to do so in such a way that the growth of US FDI in European manufacturing became part of the story. Since this has already been discussed at length in Chapter 3 it will not be elaborated upon here, save to recall that one reason for the demise of the PCM is that the technological leadership enjoyed by the USA in the 1950s and early 1960s gave way to a more balanced technological competition between the USA, Europe and Japan. Indeed, it is now Japan which has a regular trade surplus and whose firms are investing heavily in US manufacturing. Another reason is that the PCM dealt only with import-substituting investment, but since the 1970s the global integration of affiliates within MNCs has become steadily more important.

In his criticism of the PCM Kojima (1978) relied heavily on the distinction between import-substituting (trade-displacing) and offshore or export-platform (trade-creating) types of investment. An unfortunate pro-Japanese anti-American slant was given to his argument by a presumption that, through its displacing trade, import-substituting investment damaged welfare, while through its creating trade export-platform investment improved welfare; and by his labelling the former 'American-type' and the latter 'Japanese-type'. Apart from the fact that there are many US firms with export-platform investments in Southeast Asia, and many Japanese firms with import-substituting investments in Europe and the USA, there is no reason why import-substituting investments must reduce the overall extent of trade at a macroeconomic level (as opposed to the level of the individual firm) unless certain restrictive assumptions are made. Moreover, there is evidence to suggest that export-platform invesments are more likely to be of an enclave kind with little technology diffusion to host country firms (Dunning and Cantwell, 1989), and therefore they may play a lesser role in host country industrial adjustment and welfare.

Kojima's position owes much to his interpretation of FDI (unlike Vernon) within an HOS model. Import-substituting investment was therefore seen as a replacement of trade in accordance with comparative advantage, while export-platform investment involved firms in an industry comparatively disadvantaged in the home country. As an extension of this argument, Kojima and Ozawa (1985) claim that global welfare is increased where international production helps to restructure the industries of each country in line with dynamic comparative advantage.

The underlying developmental processes described by Vernon and Hirsch and by Kojima and Ozawa have many similarities despite the broader nature of the Kojima–Ozawa argument, and despite the differences in the theories of trade they employ. The Kojima–Ozawa approach applies particularly to a country which is growing rapidly, such as in recent years Japan, West Germany or the newly industrialising countries. As local firms innovate and steadily upgrade their domestic activity, they have an interest in relocating their less sophisticated types of production in countries at an earlier stage of development. Although they still possess the technological and organisational know-how to sustain this simpler production, it may be more profitable to do so abroad, and to concentrate on higher value added activities in the now more developed home environment. In other words, as in the PCM, firms relocate 'mature' or 'maturing' lines of production in countries which are still a step behind the home country.

Apart from export-platform investments, Ozawa (1979 and 1982) pays particular attention to resource-based investments. Again, in a country experiencing rapid industrial growth, and especially in a 'Ricardian' one such as Japan lacking in natural resources, the rate of domestic expansion is likely to be constrained by resource availability. Home country MNCs then have a direct interest in investing in resource-related development abroad as a means of supplying their own domestic markets. As they develop their manufacturing operations at home and in other industrialised countries, they may wish to move basic resource processing and simple manufacturing activities close to the site of resource extraction. In this case, the objectives of MNCs are in line with the development strategies of host countries.

This is an interesting idea, particularly when it is not assumed to be peculiarly true of Japanese MNCs alone, and once it is stripped from the intellectual straitjacket of the HOS model in which Kojima had confined it. Today, Japanese MNCs are at least as concerned with oligopolistic investments in the USA and Europe as with resource-based and export-platform investments in less developed countries (LDCs). It is now the newer Third World multinationals that are especially oriented to LDCs.

The major problem with Kojima's use of the HOS model is that it cannot be assumed that the industries in which a country has its greatest innovative potential are those in which it currently enjoys a comparative advantage (Pasinetti, 1981). This indeed is the lesson of Japan's own post-war history, in which she successfully encouraged local technological development in sectors in which she began with a comparative disadvantage. With the assistance of foreign MNCs through licensing agreements (Ozawa, 1974) some comparatively disadvantaged industries were transformed into comparatively advantaged ones.

Moreover, it was precisely Vernon's point that import-substituting investment need not emanate from a sector in which the home country has a comparative advantage (unlike in the HOS model). A country which holds technological leadership may be a net exporter in some comparatively disadvantaged as well as in comparatively advantaged industries, due to the technological edge of its firms. This is what gives it a regular trade surplus. If the innovative lead is lost (in the PCM as products mature) then production is relocated in accordance with underlying cost-determined comparative advantage. Kojima should also welcome this type of industrial adjustment if his argument were to be applied consistently.

A more general macroeconomic approach is the investment–development cycle, advanced by Dunning (1982), which suggests that the international direct investment position of countries relates to their stage of development. Macroeconomic developmental approaches may be used to explain longer run historical trends in the evolution of international production (Cantwell, 1988a; Cantwell and Dunning, 1984). Macroeconomic theories can also be extended to take account of the influence of financial and exchange rate factors (as emphasised by Aliber, 1970). A fuller discussion of these and other theories can be found in Cantwell (1988b).

9.2.5 Competitive International Industry Approaches

Of the four theoretical frameworks considered in Sections 9.2.2 to 9.2.5, there are two combinations which are incompatible, other than to the extent that ideas from one might be used to qualify the other. The different perspectives of the market power and internalisation theories of the firm have already been discussed; the market power and competitive international industry approaches are similarly opposed, though for slightly different reasons. Both are set out in the context of an oligopolistic industry. However, while the market power school suppose that in general internationalisation lowers the extent of competition and increases collusion amongst firms, competitive international industry approaches agree that in general the growth of international production tends to be associated with rivalry and to sustain the process of technological competition amongst MNCs. Also, the latter are genuinely mesoeconomic (industry level) approaches and do not constitute another theory of the firm.

The earliest oligopolistic theories of international production (in the rivalrous as opposed to the market power sense) were the Mark II versions of the PCM, and similar extensions of the product cycle approach. As noted in Chapter 3, Vernon had argued that cross-investment (that is, intra-industry production) developed to reduce the threat of subsidiary price cutting in the domestic market of each large firm, despite the potential cost-minimising benefits of concentrating production in just one or a few locations.

This idea of intra-industry production as an 'exchange of threats' became crucial to the work of Graham (1975, 1978 and 1985). According to Graham, oligopolistic interaction between firms in an industry increases as firms grow, since (following product cycle reasoning) the capital-intensity of production rises and economies of scale become more important as the product line matures. As the ratio of fixed to total costs rises the consequence of rivals adopting aggressive price-cutting strategies becomes potentially more damaging, and so each firm increasingly takes account of risks as well as returns. In doing so it may have to accept some trade-off between security and profitability.

The notion of the search for security is also at the heart of the market power theory of the firm, though in this case it is achieved through monopolisation and collusion and is generally supposed to be in line with profitability (Cowling and Sugden, 1987). Indeed, Hymer and Rowthorn (1970) had argued on this basis that the leading firms in each industry would aim to have a similar geographical distribution of sales or production as one another, at which point collusive agreements to ensure security would reach a peak. However, in Graham's (1975) historical account, while industrial stability and an avoidance of price warfare had been typically maintained before 1914 by collusive agreements and cartels, the exchange of threats was seen as a non-collusive alternative, which since 1960 had become the more usual means of reducing risk. With an exchange of threats competition is preserved, but in a stable rather than a cut-throat form.

To emphasise the difference with the analysis of the market power school, Graham (1985) goes beyond the product cycle framework and suggests that intra-industry production (IIP) will generally act to accelerate new product development and introduction. He therefore reaches the very opposite view to that of Hymer and Rowthorn (1970) or Cowling and Sugden (1987): 'interpenetration of national markets by MNCs based in different countries – assuming that no merger of major rivals results – acts to reduce the likelihood that collusion can be successfully undertaken globally' (Graham, 1985, p. 82).

The notion of oligopolistic interaction can also be combined with various other non-product cycle ideas on the firm. Sanna Randaccio (1980) combined oligopolistic interaction with Penrose's (1959) theory of the growth of the firm. She hypothesised that the ability of a firm to gain an increasing share of an individual market through local production was a function of the share of that market already held. Firms with a smaller existing market share would be able to grow rapidly with a lower risk of setting in motion a damaging competitive warfare. To allow a stable competitive process to continue to run smoothly, it was therefore in the interests of US firms to switch resources away from domestic growth and towards the expansion of their European production (starting from a smaller base), just as it was in the interests of European firms to expand their US operations.

Meanwhile, returning to the exchange of threat view of oligopolistic interaction, Casson (1987) has attempted to integrate this with the internalisation theory of the firm. In this case, in order to examine the implications for rivalry of different relative market shares, Casson's model makes the determinant of long-run market share exogenous. Depending upon the relative strength of firms, an exchange of threats (establishing a foothold in the major markets of rivals) may be used to preserve price stability.

This book has attempted to switch the emphasis towards technological competition between the MNCs in an international industry, rather than on the effect of oligopolistic rivalry on price competition. The quotation from Graham above suggests a connection between technological competition and the rise of IIP, which is supported by the evidence of Chapter 7. There is good reason to argue that the increasing internationalisation of manufacturing production has helped to sustain technological competition between MNCs. The development of technology and the growth of international production have proceeded as related cumulative processes. The other arguments of earlier chapters need not be reiterated here.

Some similar themes appear in the internationalisation of capital approach of Jenkins (1984 and 1987). In his view the growth of international production is just one aspect of a trend towards a more integrated world economy. As a result of this trend in each industry products and processes have become increasingly standardised across countries, while firms safeguard their competitive positions through the continuing differentiation of products and technology. Once again, the growth of MNCs is seen as part of a competitive process, in which each firm attempts to gain competitive advantages through innovation, and only in certain circumstances do they enter into (for a time) collusive arrangements.

According to Jenkins, if nationalistic governments in LDCs prevent the entry of MNCs this only reduces the speed of adaptation to the requirements of international competition, but does not allow them to avoid such adaptation altogether. Marxists such as Warren (1980) go further and argue that FDI helps to promote local capitalist development and economic advance. They emphasise oligopolistic competition between MNCs of different national origins. They reject the idea that MNCs provide host LDCs with inappropriate technology, and contend that this is simply a defence of economic backwardness (Emmanuel, 1982).

The various approaches to modelling the growth of international production differ over what they implicitly suppose to be the driving force underlying this expansion. In the technological accumulation approach it is the conditions for technology creation and its effective and efficient use in production. In the market power approach it is the widening of collusive networks and the restriction of competition in each national market. For internalisation theorists whether international production expands or contracts depends upon changes in the transaction costs of operating in a wider set of markets (including the market for technology), relative to the costs of the direct coordination of transactions. In most macroeconomic approaches it is the developmental position of countries and their firms. For theorists who focus on the foreign direct investment flows associated with international production rather than international production itself (such as Aliber, 1970), it is the functioning of currency and financial markets. For business economists whose method is to provide a series of case studies of MNCs (such as Stopford and Turner, 1986), the driving force is individual managerial or entrepreneurial strategies. These approaches are not always mutually exclusive, but Section 9.3 examines some points at issue between them in the context of theorising about the dynamic or cumulative aspects of international production.

9.3   Two Points at Issue between the Different Theories

There are two major points at issue between the different theoretical approaches to international production which affect the way in which they analyse the growth of international production. Using the terminology of the eclectic paradigm, the first concerns the interpretation to be placed on ownership advantages, while the second is the significance attached to the interaction between ownership and location advantages (and/or internalisation and location advantages). In this section these points are considered in turn.

In the evolutionary approach of earlier chapters, the starting point was the dynamic one of the creation and accumulation of technology and capital by groups of active firms. International production is then essentially explained as the developmental outcome of the competitive process between firms belonging to common industrial groups. Where an industrial group of firms in a given country generate a particular path of technological development, they may establish an innovative centre in which production becomes locationally advantaged and in which a specific technological tradition is created.

They are also enabled to spread their productive networks across other locationally advantaged sites. Market growth as well as innovation is likely to be favoured in such locations. This view relies crucially on the creation of a continuous stream of ownership advantages in a process of technological competition, and on a dynamic interaction between the ownership advantage of groups of firms and the locational advantages of the sites in which they produce. MNCs in manufacturing emerged historically and became successful through the generation and the cultivation of innovative ownership advantages.

Ownership advantages are defined by reference to the (oligopolistic) final product market. They are advantages which lower the unit costs and raise the profit margins of given firms relative to others in the same industry. A firm with weak ownership advantages (and certainly one with none at all) has high unit costs relative to others in its industry, and consequently suffers consistent losses. A firm with strong ownership advantages (measured, for example, by a high share of patenting activity) will have a larger market share, and if its ownership advantages become stronger still (its share of patenting rises) then its market share rises. Firms which lose ownership advantages face lower profits and a reduction in their market share. Firms or MNCs therefore require ownership advantages as a necessary condition of their continued existence, and the stronger are their advantages the faster the rate of international growth that they are able to sustain.

In the technological accumulation approach each firm in an industry internally generates a succession of innovations or ownership advantages, which constitute the basis of both their domestic and their international production. The ability to innovate in a growing sector is itself an ownership advantage which is a function of the past technological experience of the firm. These ownership advantages are oligopolistic advantages, essential to survival in an international oligopoly. The most innovative firms in an industry create a faster stream of more effective ownership advantages, and in so doing they increase their international production more rapidly and raise their world market shares. The weakest firms lose market shares, and if a firm lost all ownership advantages it would be quickly driven out of both domestic and international markets. However, firms with the fewest or weakest ownership advantages in general hold their position more easily in domestic markets than in international markets due to government support, consumer loyalty, the closeness of local business contacts and so forth.

Against this, it is sometimes claimed by writers in the internalisation tradition that ownership advantages are not necessary for the existence of MNCs and international production, apart from the advantages created by internalisation itself (Buckley and Casson, 1976; Casson, 1987). By this they mean that firms may grow relative to intermediate product markets where producers trading with one another merge their operations, irrespective of the ownership advantages of the productive units concerned. Hence they are referring strictly to the transition from national firms combining to form multinational firms, by means of the internalisation of the intermediate product markets which previously linked them.

However, to conclude from this that ownership advantages are not a necessary condition for international production is a very misleading way of putting matters. At least one of the individual productive units that combine to form an MNC must have ownership advantages, so that the MNC does too. For firms to grow relative to their competitors in final product markets, or indeed simply to retain their share of final product markets, ownership advantages are necessary. The generation of ownership advantages, achieved mainly through innovation, is necessary for competitive success and indeed survival. Firms that fail to accumulate technology and related ownership advantages are either driven out of business, or taken over by firms that have the capacity to do so.

Internalisation may complement ownership advantages but it is not a substitute for them. A group of firms each of which has no ownership advantages and therefore makes losses due to the high level of production costs will not suddenly become profitable merely by organising transactions among themselves more effectively (lowering transaction costs). They will only reduce their losses. In any event, internalisation advantages will tend to be greater where each participating affiliate begins with strong ownership advantages, due to the potential for economies of scope, technological complementarities, and so forth.

The need for ownership advantages applies both to national firms and MNCs, although it is especially true of MNCs, since inefficient firms may be able to survive longer in their own domestic markets. Moreover, the most innovative firms whose ownership advantages are strongest are more likely to take over firms whose ownership advantages are weak than vice versa. The international networks of MNCs expand directly through the independent establishment and extension of their own ventures, together with takeovers or mergers with firms whose technological activity (ownership advantages) is complementary. Once MNCs are created they may gain additional advantages from the international coordination of activity, in part through the enhancement of their technological strengths by way of a more geographically and industrially diversified research programme. The international coordination of research and production is therefore supportive of technological advantages, not a substitute for them, nor a sufficient condition for their generation.

The debate over the necessity or otherwise of ownership advantages is largely a discussion that has proceeded at cross-purposes, as different approaches have been set up to answer different questions. Internalisation theory sets out to explain why firms in general displace intermediate product markets in general, or why MNCs in general displace international trade in intermediate products, in which it is not necessary to refer to ownership advantages. The existence of ownership advantages on the part of firms can be subsumed in the Coasian theory under the existence of firms themselves, the joint ownership of assets being explained by the replacement of markets.

However, while this is a sufficient theory of the growth of firms considered as a whole (in place of atomistic competition), it is not a sufficient theory of the growth of a particular firm or a particular group of firms *vis-à-vis* other firms. To explain why one firm displaces another, or why for example Japanese MNCs have grown at the expense of US firms, then relatively stronger ownership advantages are necessary. The relative strength of technological ownership advantages is in general the principal determinant of variations in unit costs or productivity across firms (particularly in industrialised countries). Variations in unit costs are in turn necessary to explain why certain firms grow faster than their competitors in a given final product market.

In this respect, internalisation theory has addressed a different issue to competitive international industry approaches; it is concerned with the extent of firms as a whole as opposed to intermediate product markets as a whole, rather than the process of competition between already existing firms. For a particular firm or a particular group of firms ownership advantages are a necessary condition for establishing and preserving international production.

A similar criticism can be found in Kojima's work on Japanese MNCs and the recent literature on Third World MNCs, in which it is sometimes claimed that investments in LDCs require no ownership advantages on the part of the firms that make them. It is certainly true that it may be the smaller relatively weaker firms in an international industry which make export-platform investments in labour-intensive production, while the more advanced firms invest in research-related production in the industrialised countries. However, this need not be the case in the Kojima–Ozawa story, since it may be the same firms which upgrade their production at home while redeploying less sophisticated types of activity in countries at an earlier stage of development. Even if it is the case all that this entails is that the ownership advantages of such smaller newer MNCs lie in a simpler type of manufacturing and in their access to commercial outlets, not that they have no ownership advantages at all. It is simply that the more sophisticated types of international production require stronger ownership advantages (a stronger technological base) than other types, which is scarcely surprising.

Note also that ownership advantages are not always necessary in the specific sector in which investment takes place. Consider, for example, a diversified takeover by an international conglomerate. Ownership advantages are therefore to be understood as providing a firm with a general capability to expand, and the stronger are these advantages the faster the speed of international expansion that it is able to sustain. It is true that where a conglomerate chooses to move into an unrelated sector its motive is essentially of a financial kind, and may be better explained by a theory of FDI (such as that of Aliber, 1970, or Rugman, 1979) than a theory of international production. The ownership advantages of such firms may well be of a financial rather than a technological kind.

Where firms rely on technological ownership advantages they invest in production in areas related to their existing strengths. Of course, where they produce in foreign centres of innovation they may also gain access to a new stream of complementary technological developments. Investment may be motivated mainly by the objective of strengthening their ownership advantages. In such cases the original ownership advantages of firms are built upon, consolidated in their existing fields and extended into new areas. There is a progressive interaction between the growth of international networks and the strength of ownership advantages.

The other confusion about ownership advantages that is often found in the literature is that, as noted in Section 9.2.1, they are frequently described as monopolistic advantages. Perhaps one reason why internalisation theorists have objected to the concept of ownership advantages in the eclectic paradigm is because of this interpretation, associated with the market power approach. The origins of this terminology go back to Kindleberger (1969), who recast Hymer's work to associate MNCs with the existence of a particular market structure, that of monopolistic competition (within which each participating firm has some monopolistic advantage). Hymer himself believed that firms actively sought to raise barriers to entry and to collude with other firms in their industry, and that market structure was a product of their behaviour and the extent to which they succeeded in making lasting collusive agreements, rather than the other way around. In Kindleberger's restatement the MNC was seen as a function of a market structure characterised by monopolistic competition between differentiated products, rather than as an active agent engaged in oligopolistic interaction with other firms.

In Hymer's approach the strength of ownership advantages amongst the leading firms reflects the degree of monopoly power in a sector. However, strictly speaking even in his framework ownerhip advantages should be thought of as oligopolistic and not monopolistic advantages.[4] This is even more true in competitive international industry approaches such as that based on technological accumulation. Firms accumulate differentiated but overlapping technologies, irrespective of whether they produce identical or different final products. MNCs are therefore involved in technological competition with other members of an international oligopoly. Where it suits them they may arrive at certain cooperative arrangements, such as in cross-licensing agreements. Each participant in the international oligopoly has certain specific advantages based on their own previous technological experience. The relative strength of the ownership advantages of each firm determines both their rate of growth *vis-à-vis* their competitors, and their attitude towards cooperative arrangements with other firms in the same sector.

The second point at issue between approaches is whether or not they make allowance for an interaction between ownership (or internalisation) and location advantages. The dividing line here is essentially between the two theories of the firm, which tend to treat location as exogenous (to be determined by some other theory), and the macroeconomic and mesoeconomic approaches in which locational factors are themselves influenced by the growth of firms. To give an illustration of a process of continual interaction between ownership and location advantages, consider the case of a modern global manufacturing industry (as described in Chapters 7 and 8) in which MNCs require direct access to all the main locationally advantaged centres of technological development. By investing in research and production in these sites they increase their own competitiveness (ownership advantages), as well as the competitiveness of production in the countries concerned and their attractiveness to other firms (location advantages). In this case ownership and location advantages are not independent.

Looking back historically, the location advantages of countries also helped to determine the pattern of technological specialisation of each country's firms, or in other words the sectors in which they had the greatest capacity for generating ownership advantages. For example, Rosenberg (1976) has shown that US firms emerged from the industrial revolution with advantages in woodworking technologies due to the plentiful availability of local timber supplies, while British firms developed advantages in metal working and coal-related technologies due to local coal deposits.

Internalisation theorists have tended to take location advantages and technological ownership advantages as given and exogenous, in order to focus attention on the form of linkages between plants. Neoclassical trade and location theorists have gone even further, and in many cases simply assumed away technological ownership advantages (all firms have access to the same technology) in order to focus on purely locational factors. The market power school treat technological advantages as barriers to entry, while typically taking location as exogenous.

This is not to say that the market power and internalisation approaches have ignored locational factors. Hymer (1975), for example, suggests that firms locate activities in accordance with a hierarchy, with high-grade activities in industrialised countries and low-grade activities in LDCs. Newfarmer (ed., 1985) and Bornschier and Chase-Dunn (1985), who take a market power view, similarly distinguish between central and peripheral locations in an international industry. While retaining the internalisation perspective, Casson et al. (1986) have also suggested model of the international division of labour. Indeed, internalisation theorists have recognised that the growth of the firm may influence location; for example: 'internalisation allows international transfer price manipulation that will bias location towards inclusion of low tax locations' (Buckley, 1988, p. 182). It is simply that those who have analysed the growth of international production from the viewpoint of the theory of the firm have not on the whole attached very much significance to the interaction between the growth of the firm and the changing location of production.

This contrasts with the product cycle approach, the Kojima–Ozawa analysis and the various competitive international oligopoly theories, which have all stressed such interaction and the consequences for the structure of industrial production and trade. Macroeconomic theories of international production make location advantages depend on macroeconomic factors related to countries and their level of development, while mesoeconomic approaches emphasise the locational factors specific to an international industry (see Gray, 1982, for a discussion of whether location advantages can be treated as purely macroeconomic). Naturally, the focus of interest and the level at which analysis is conducted affect the treatment of location in theories of international production.

9.4   *The Development of an Evolutionary Analysis of International Production*

As a rule, macroeconomic and competitive international industry approaches to international production are by their very nature dynamic or evolutionary, as they have usually been concerned to describe a process over time. Theories of the firm, although they have attracted a greater literature, have addressed only certain types of question on the dynamics of international production. The internalisation approach has asked why firms in general (the visible hand) have expanded relative to markets in general (the invisible hand). The particular issue which has attracted most attention is the idea of an evolution in the business of firms from exports (through a sales agent), to licensing a foreign company, to the establishment of international production (Buckley and Casson, 1981; Teece, 1983; Nicholas, 1986). This section considers how theories of the firm may be extended or adapted to take account of the wider evolution of international production linked to the cumulative development of technology.

Two possible extensions of internalisation theory might be considered which would broaden the approach. Firstly, attention can be given to the interrelationship between the growth of the firm and the changing location of production, which has already been raised in the context of other approaches as suggested in the previous section. Secondly, the transaction cost framework might be usefully combined with a theory of entrepreneurship, innovation or the changing technology and organisation of production within the firm.

Whereas the theory of cumulative technological change, like the work of the classical economists, is a theory of production (and the changing technology of production), as it stands the Coasian theory of the firm, like the neoclassical economic thinking of which it is a criticism, is a pure theory of exchange. Exchange takes place under a variety of institutional arrangements, in markets or within the firm. It is worth noting that Nelson and Winter (1982) make use of certain aspects of the work of Williamson and others in developing their evolutionary theory of economic change. However, in order to make the Coasian theory of the firm itself evolutionary it would be necessary to specify how transaction costs are themselves influenced by the growth and technological innovation of firms.

As Casson (1986) has pointed out, while transaction cost theory specifies conditions under which non-market institutional arrangements will obtain (for example, within the firm), it does so at present to the exclusion of any active role for managerial strategy. In other words, while a theory of the MNC, couched in an exchange framework, can explain the existence of the MNC or the firm, it has still left the firm itself as a passive reactor to transactional circumstances. Changes in the organisation or control of production are merely a response to changes in the costs of various exchange relationships in markets or otherwise. This procedure may be justified if one is concerned only with the internalisation of intermediate product markets actively replacing trade between independent parties.

However, difficulties are encountered when this approach is applied to the historical evolution of say, import-substituting international production; take, for example, the growth of investment by US firms in Europe in the 1950s and 1960s. Part of the reason is that locational factors were responsible (faster growth in Europe). While this helps to explain the initial direction of the investment, it does not explain how manufacturing MNCs steadily grew in the post-war period. Nor does a change in political risk provide a complete explanation. It might be suggested that where markets in intangible assets such as technology become in some sense more imperfect, the transaction costs of market exchange rise, and horizontally integrated MNCs tend to grow. But it seems most implausible to argue that any international markets were operating more imperfectly in the late 1950s and the 1960s by comparison with the 1930s.

Within this framework, the explanation is presumably in part that the transaction costs of cooperative relationships fell even faster than the costs of using the market mechanism. Yet one of the main reasons for this was the growing experience of firms in international production in the industrialised countries, initially of US firms in Europe. To avoid falling into a static tautology (that is, that international transaction costs fell because of the growth of the MNC, and the MNC grew because of lower internal transaction costs) then this must be set in a dynamic framework, whereby the interaction between transaction costs and the growth of firms is set out. Transaction costs depend upon the technological and productive activity of the firm. The international growth of firms is part of an evolutionary process. This also helps to call attention to other aspects of the explanation: in this case, the generation of strong technological ownership advantages by US firms, and the greater locational advantage of producing in Europe in the 1960s.

Internalisation theory when taken independently is at its strongest when discussing vertical integration, such as the backward integration of manufacturing companies to secure raw material supplies including natural resources (Casson et al., 1986), or their forward integration into distribution or sales agencies (Nicholas, 1983 and 1985). When it comes to the international expansion of manufacturing production itself a pure theory of exchange is on weaker ground. Technology may accumulate within the firm not so much because of the characteristics of the market for technology once it has been created, as because of the conditions under which it is most easily generated and used in production. Technology is then difficult and costly to transfer or exchange between firms precisely because it tends to be associated with the research and production experience accumulated within a particular firm, rather than vice versa. Armour and Teece (1980) take a step in this direction in recognising from a transaction costs perspective that vertical integration may increase R&D and the rate of technological change.

It seems reasonable to suppose that the accumulation of technology and the growth of production within the firm will affect the transaction costs of exchange. The transaction cost theorist has instead tended to start from exchange in a market, which gives way to more consciously organised control where it is relatively inefficient. Coase (1937) stressed the market conditions which lead to a reorganisation or an extension in the organisation of the firm. However, the nature and extent of a firm's transactions and cooperative arrangements with other firms, as well as its market share, also depend upon its innovative capacity *vis-à-vis* other firms.

A similar criticism can be made of the conventional structure–conduct–performance paradigm of industrial organisation theory which is essentially static and does not deal with industrial dynamics or evolution (Carlsson, 1987). Once again, the firm appears as a passive reactor to changes in market structure. This is rather unfortunate, since early analysis of the growth of the firm in the 1950s (such as Penrose, 1959) had appeared to fit in nicely with treatments of imperfect competition, and the emergence of the study of industrial economics. Indeed, as noted above, Hymer's theory of the growth of the MNC was based precisely on this combination of an active firm increasing the extent of its market power, and colluding with others to raise what Bain (1956) termed barriers to entry or new competition.

However, there are problems with the market power theory of the firm if it is used as a general explanation of the growth of international production, especially in the context of the technological competition that exists today between the firms of the major industrialised countries. The implication of the market power view is that profitability is usually seen as being raised through a restriction of output and of the number of firms competing. In fact the steady generation of ownership advantages through technological accumulation, rather than serving as barriers to potential new entrants may increase competition amongst firms already in the international oligopoly, causing them to expand their own research and production. Research-intensive lines of activity are not generally known for their lack of competition in world markets, but quite the reverse.

In an industry with rapid technological accumulation the high profitability associated with the existence of strong ownership advantages is not due to a restriction of output, a lack of competition, and a high degree of industrial concentration as a feature of market structure. It is rather accounted for by the greater scope that innovation provides for productivity improvements, and with them the faster growth of output. Competition is more intense in terms of the creation of new products and processes.

Yet the market power theory of the firm as used by Hymer may be relevant to certain types of MNC activity, even if it does not provide a general explanation of MNC growth. Firms are more likely to seek out cooperative agreements or collusion with other firms where they are in a weak position *vis-à-vis* other firms in their industry, or where competition is very intense. Where they are relatively weak the licensing of technology is liable to run in just one direction, while where they are a leading member of an international oligopoly cross-licensing arrangements are to be expected. Collusion between MNCs is quite possible, but it is generally to improve the relationship between their international production networks rather than being an explanation of them.

Analysing trends in technological competition (and cooperation) between MNCs in the future is likely to be of considerable interest. The question perhaps most directly raised by the evidence of this book is the extent to which the recent upsurge in the internationalisation of production is leading to an increasing locational concentration of research activity. This in turn has effects on the competitiveness of both firms and countries. Work on international industrial dynamics of this kind is still at a relatively early stage, but similar work on cumulative patterns of technological change is already attracting increasing attention in other fields. Further explorations of the relationship between technological innovation and the production of MNCs hopefully will make a distinctive and important contribution.

This book has examined the evolution of international manufacturing industries in the post-war period, and particularly the evolution of technological competition. It is to be hoped that the usefulness of the theory of technological accumulation and of an evolutionary framework of analysis in this field has been demonstrated. The evolutionary and cumulative approach to the organisation of technology and production gives a new orientation to the theory of international production, as shown in this chapter, and to the theory of international trade. It suggests that further explorations of the relationship between technological innovation and the production of MNCs will help to deepen economists' understanding of the modern development of international economic activity.

NOTES

1 Dunning (1983 and 1988a) refers to these as ownership advantages of an asset kind, and ownership advantages of a transaction cost minimising kind, or, following Teece (1983), ownership advantages of a governance cost minimising kind.
2 Although the distinction between appropriability and coordination remains an important conceptual one, in practice where the firm controls complementary assets it may be difficult to differentiate between the two. The significance of this division is discussed further by Teece (1989).
3 That Cowling and Sugden do not seem to be aware that they are writing in the Hymer tradition need not be of concern here.

4 They are monopolistic only if they are based entirely on product differentiation. If firms have advantages that are related to production – scale economies, patented technology, or high start-up costs – then they need not have monopolistic control of (segments of) the final product market. Their monopoly power is a result of collusion, which is made easier by such barriers to entry, but is exercised jointly rather than individually. It might still be said that each firm has a quasi-monopolistic position in the everyday sense that each exercises a monopoly over the use of its own patented technology or large-scale plants (Jenkins, 1987). However, if they do not sell the use of such technology or plants in an external market then they are not monopoly sellers; hence the term *quasi*-monopoly. The market they actually serve is oligopolistic, even if it is divided by collusive agreement.

References

Acocella, N. (ed., 1985), *Le Imprese Multinazionali Italiane*, Bologna: Il Mulino.

Aliber, R.Z. (1970), 'A theory of direct foreign investment', in Kindleberger, C.P. (ed.), *The International Corporation: A Symposium*, Cambridge, Mass.: MIT Press.

Aquino, A. (1978), 'Intra-industry trade and intra-industry specialisation as concurrent sources of international trade in manufactures', *Weltwirtschaftliches Archiv*, vol. 114, no. 2.

Archibugi, D. (1986), 'Sectoral patterns of industrial innovation in Italy', Consiglio Nazionale delle Ricerche Discussion Paper.

Armour, H.O. and Teece, D.J. (1980), 'Vertical integration and technological innovation', *Review of Economics and Statistics*, vol. 62.

Arthur, W.B. (1984), 'Industry location patterns and the importance of history', Stanford University Center for Economic Policy Research Discussion Paper.

Arthur, W.B. (1988), 'Competing technologies: an overview', in Dosi, G., Freeman, C. Nelson, R.R., Silverberg, G. and Soete, L.L.G. (eds), *Technical Change and Economic Theory*, London: Frances Pinter.

Arthur, W.B., Ermoliev, Y.M. and Kaniovski, Y.M. (1987), 'Path-dependent processes and the emergence of macro-structure', *European Journal of Operational Research*, vol. 30.

Atkinson, A.B. and Stiglitz, J.E. (1969), 'A new view of technological change', *Economic Journal*, vol. 79, no. 3, September.

Baba, Y. (1987), 'Internationalisation and technical change in Japanese electronics firms, or why the product cycle doesn't work'. Paper presented at EIASM meeting on Internationalisation and Competition, Brussels, June.

Bain, J.S. (1956), *Barriers to New Competition*, Cambridge, Mass.: Harvard University Press.

Balassa, B. (1965), 'Trade liberalisation and "revealed" comparative advantage', *The Manchester School*, vol. 33, no. 2, May.

Baran, P.A. and Sweezy, P.M. (1966), *Monopoly Capital*, New York: Monthly Review Press.

Bergsten, C.F., Horst, T. and Moran, T.H. (1978), *American Multinationals and American Interests*, Washington: The Brookings Institution.

Bornschier, V. and Chase-Dunn, C. (1985), *Transnational Corporations and Underdevelopment*, London: Greenwood.

Brech, M. and Sharp, M. (1984), *Inward Investment: Policy Options for the United Kingdom*, London: Routledge and Kegan Paul.

Buckley, P.J. (1983), 'New theories of international business: some unresolved issues', in Casson, M.C. (ed.), *The Growth of International Business*, London: Allen and Unwin.

Buckley, P.J. (1985), 'The economic analysis of the multinational enterprise: Reading versus Japan?', *Hitotsubashi Journal of Economics*, vol. 26, no. 2, December.

Buckley, P.J. (1988), 'The limits of explanation: testing the internalisation theory of the multinational enterprise', *Journal of International Business Studies*, vol. 19, no. 2, summer.

Buckley, P.J. and Casson, M.C. (1976), *The Future of the Multinational Enterprise*, London: Macmillan.

Buckley, P.J. and Casson, M.C. (1981), 'The optimal timing of a foreign direct investment', *Economic Journal*, vol. 91, no. 1.

Buckley, P.J. and Casson, M.C. (1985), *The Economic Theory of the Multinational Enterprise: Selected Papers*, London: Macmillan, and New York: St. Martin's Press.

Buckley, P.J. and Roberts, B.R. (1982), *European Direct Investment in the USA Before World War I*, London: Macmillan.

Burstall, M., Dunning, J.H. and Lake, A. (1981), *Multinational Corporations, Governments and Technology: The Pharmaceutical Industry*, Paris: OECD, spring.

Cantwell, J.A. (1984), 'The revaluation of the UK's outward foreign direct investment stock', mimeo, University of Reading.

Cantwell, J.A. (1986a), Review of Helpman, E. and Krugman, P.R., 'Market Structure and Foreign Trade: Increasing Returns, Imperfect Competition, and the International Economy', *Economic Journal*, vol. 96, no. 1, March.

Cantwell, J.A. (1986b), 'Recent trends in foreign direct investment in Africa', in Cable, V. (ed.), *Foreign Investment Policies and Prospects in Africa*, London: Commonwealth Secretariat.

Cantwell, J.A. (1986c), 'Technological Innovation and International Production in the Industrialised World: A Study of the Accumulation of Technology and Capital in International Networks'. Unpublished Ph.D. thesis, University of Reading.

Cantwell, J.A. (1987), 'The reorganisation of European industries after integration: selected evidence on the role of multinational enterprise activities', *Journal of Common Market Studies*, vol. 26, no. 2, December; reprinted in Dunning, J.H. and Robson, P. (eds), *Multinationals and the European Community*, Oxford: Basil Blackwell.

Cantwell, J.A. (1988a), 'The changing form of multinational enterprise expansion in the twentieth century', in Teichova, A., Lévy-Leboyer, M. and Nussbaum, H. (eds), *Historical Studies in International Corporate Business*, Cambridge: Cambridge University Press.

Cantwell, J.A. (1988b), 'Theories of international production', University of Reading Discussion Paper in International Investment and Business Studies, no. 122, September.

Cantwell, J.A. (1989), 'The growing internationalisation of industry: a comparison of the changing structure of company activity in the major industrialised countries', in Dunning, J.H. and Webster, A.D. (eds), *Structural Change through the World Economy*, London: Routledge.

Cantwell, J.A. and Dunning, J.H. (1984), 'The emergence of multinationals in the organisation of international production', in Fonseca, A. (ed.), *Multinationals in Third World Countries: towards a code of conduct*, Rome: IFCU Centre of Research.

Cantwell, J.A. and Dunning, J.H. (1985), 'The "New Forms" of International Involvement of British Firms in the Third World', Report prepared for the OECD, January.

Cantwell, J.A. and Tolentino, P.E.E. (1987), 'Technological accumulation and Third World multinationals'. Paper presented at the Annual Meeting of the European International Business Association, Antwerp, December.

Carlsson, B. (1987), 'Reflections on "industrial dynamics": the challenges ahead', *International Journal of Industrial Organisation*, vol. 5, no. 2, June.

Casson, M.C. (1981), Foreword, in Rugman, A.M., *Inside the Multinationals: the Economics of Internal Markets*, London: Croom Helm.

Casson, M.C. (1982), 'The theory of foreign direct investment', in Black, J. and Dunning, J.H. (eds), *International Capital Movements*, London: Macmillan.

Casson, M.C. (1983), 'Introduction: the conceptual framework' in Casson, M.C. (ed.), *The Growth of International Business*, London: Allen and Unwin.

Casson, M.C. (1986), 'General theories of the multinational enterprise: a critical examination', in Jones, G. and Hertner, P. (eds), *Multinationals: Theory and History*, Farnborough: Gower.

Casson, M.C. (1987), *The Firm and the Market: Studies in Multinational Enterprise and the Scope of the Firm*, Oxford: Basil Blackwell.

Casson, M.C. with Barry, D., Foreman-Peck, J., Hennart, J.-F., Horner, D., Read, R.A. and Wolf, B.M. (1986), *Multinationals and World Trade: Vertical Integration and the Division of Labour in World*

Industries, London: Allen and Unwin.

Caves, R.E. (1974), 'Multinational firms, competition, and productivity in host-country industries', *Economica*, vol. 41, no. 2, May.

Caves, R.E. (1980), 'Investment and location policies of multinational companies', *Schweiz, Zeitschrift fur Volkwirtschaft und Statistik*, vol. 116, no. 3.

Caves, R.E. (1982), *Multinational Enterprise and Economic Analysis*, Cambridge: Cambridge University Press.

Caves, R.E. (1986), Review of Erdilek, A. (ed., 1985) in *Journal of International Business Studies*, vol. 17, no. 2, Summer.

Clegg, L.J. (1987), *Multinational Enterprise and World Competition: A Comparative Study of the USA, Japan, the UK, Sweden, and West Germany*, London: Macmillan.

Coase, R.H. (1937), 'The nature of the firm', *Economica*, vol. 4, no. 4, November.

Cowling, K. (1986), 'The internationalisation of production and de-industrialisation', in Amin, A. and Goddard, J. (eds), *Technological Change, Industrial Restructuring and Regional Development*, London: Allen and Unwin.

Cowling, K. and Sugden, R. (1987), *Transnational Monopoly Capitalism*, Brighton: Wheatsheaf.

Creedy, J. (1985), *Dynamics of Income Distribution*, Oxford: Basil Blackwell.

Dasgupta, P. (1987), 'The economic theory of technology policy: an introduction', in Dasgupta, P. and Stoneman, P. (eds), *Economic Policy and Technological Performance*, Cambridge: Cambridge University Press.

Dosi, G. (1984), *Technical Change and Industrial Transformation*, London: Macmillan.

Dosi, G., Freeman, C., Nelson, R.R., Silverberg, G. and Soete, L.L.G. (eds, 1988), *Technical Change and Economic Theory*, London: Frances Pinter.

Doz, Y. (1986), *Multinational Strategic Management: Economic and Political Imperatives*, Oxford: Pergamon Press.

Dunning, J.H. (1958), *American Investment in British Manufacturing Industry*, London: Allen and Unwin.

Dunning, J.H. (1970a), *Studies in International Investment*, London: Allen and Unwin.

Dunning, J.H. (1970b), 'Technology, United States investment, and European economic growth', in Kindleberger, C.P. (ed.), *The International Corporation: A Symposium*, Cambridge, Mass.: MIT Press.

Dunning, J.H. (1971), 'United States foreign invesment and the technological gap', in Kindleberger, C.P. and Shonfield, A. (eds), *North American and Western European Economic Policies*, London: Macmillan.

Dunning, J.H. (1972), *The Location of International Firms in an Enlarged EEC: An Explanatory Paper*, Manchester: Manchester Statistical Society Occasional Paper.

Dunning, J.H. (1974), 'Multinational enterprises, market structure, economic power and industrial policy', *Journal of World Trade Law*, vol. 8, no. 6, November–December.

Dunning, J.H. (1977), 'Trade, location of economic activity, and the MNE: a search for an eclectic approach', in Ohlin, B., Hesselborn, P.-O. and Wijkman, P.M. (eds), *The International Allocation of Economic Activity*, London: Macmillan.

Dunning, J.H. (1980), 'A note on intra-industry foreign direct investment', *Banca Nazionale del Lavoro Quarterly Review*, no. 139, December.

Dunning, J.H. (1981), *International Production and the Multinational Enterprise*, London: Allen and Unwin.

Dunning, J.H. (1982), 'Explaining the international direct investment position of countries: towards a dynamic or developmental approach', in Black, J. and Dunning, J.H. (eds), *International Capital Movements*, London: Macmillan.

Dunning, J.H. (1983), 'Market power of the firm and international transfer of technology', *International Journal of Industrial Organisation*, vol. 1, no. 1, December.

Dunning, J.H. (ed., 1985), *Multinational Enterprises, Economic Structure and International Competitiveness*, Chichester: John Wiley and Sons.

Dunning, J.H. (1986), *Japanese Participation in UK Manufacturing Industry*, London: Croom Helm.

Dunning, J.H. (1988a), 'The eclectic paradigm of international production: an update and a reply to its critics', *Journal of International Business Studies*, vol. 19, no. 1, Spring.

Dunning, J.H. (1988b), 'The theory of international production', *The International Trade Journal*, vol. 3, Fall.

Dunning, J.H. and Cantwell, J.A. (1982), 'Inward direct investment from the US and Europe's technological competitiveness', University of Reading Discussion Paper in International Invesment and Business Studies, no. 65.

Dunning, J.H. and Cantwell, J.A. (1987), *The IRM Directory of Statistics of International Investment and Production*, London: Macmillan, and New York: New York University Press.

Dunning, J.H. and Cantwell, J.A. (1989), 'The changing role of multinational enterprises in the international creation, transfer and diffusion of technology', in Arcangeli, F., David, P.A., and Dosi, G. (eds), *Technology Diffusion and Economic Growth: International and National Policy Perspectives*, Oxford: Oxford University Press.

Dunning, J.H. and Norman, G. (1986), 'Intra-industry investment', in Gray, H.P. (ed.), *Uncle Sam as Host*, Greenwich, Conn.: JAI Press.

Dunning, J.H. and Pearce, R.D. (1985), *The World's Largest Industrial Enterprises, 1962–1983*, Farnborough: Gower.

Dunning, J.H., Cantwell, J.A. and Corley, T.A.B. (1986), 'An exploration of some historical antecedents to the modern theory of international production', in Jones, G. and Hertner, P. (eds), *Multinationals: Theory and History*, Farnborough: Gower.

Emmanuel, A. (1982), *Appropriate or Underdeveloped Technology?*, Chichester: John Wiley and Sons.

Erdilek, A. (ed., 1985), *Multinationals as Mutual Invaders: Intra-Industry Direct Foreign Investment*, London: Croom Helm.

Fishwick, F. (1982), *Multinational Companies and Economic Concentration in Europe*, Farnborough: Gower.

Flaherty, M.T. (1983), 'Market share, technology leadership, and competition in international semiconductor markets', in Rosenbloom, R.S. (ed.), *Research on Technological Innovation, Management and Policy*, vol. 1, Greenwich, Conn.: JAI Press.

Flowers, E.B. (1976), 'Oligopolistic reaction in European and Canadian direct investment in the US', *Journal of International Business Studies*, vol. 7, no. 2, fall/winter.

Foreman-Peck, J. (1986), 'The motor industry', in Casson, M.C. et al. (eds), *Multinationals and World Trade: Vertical Integration and the Division of Labour in World Industries*, London: Allen and Unwin.

Franko, L.G. (1976), *The European Multinationals*, London: Harper-and Row.

Freeman, C. and Perez, C. (1989), 'The diffusion of technical innovations and changes of techno-economic paradigm', in Arcangeli, F., David, P.A. and Dosi, G. (eds), *Modern Patterns in Introducing and Adopting Innovations*, Oxford: Oxford University Press.

Freeman, C. and Soete, L.L.G. (1989), 'Innovation diffusion and employment policies', in Arcangeli, F., David, P.A. and Dosi, G. (eds), *Technology Diffusion and Economic Growth: International and National Policy Perspectives*, Oxford: Oxford University Press.

Freeman, C., Clark, J. and Soete, L.L.G. (1982), *Unemployment and Technical Innovation*, London: Frances Pinter.

Galton, F. (1889), *Natural Inheritance*, London: Macmillan.

Giddy, I.H. (1978), 'The demise of the product cycle model in international business theory', *Columbia Journal of World Business*, vol. 13, no. 1, Spring.

Globerman, S. (1979), 'A note on foreign ownership and market structure in the United Kingdom', *Applied Economics*, vol. 11, no. 1, March.

Goodwin, R.M. (1967), 'A growth cycle', in Feinstein, C.H. (ed.), *Capi-

talism and Economic Growth, Cambridge: Cambridge University Press.

Goodwin, R.M. and Punzo, L.F. (1987), *The Dynamics of a Capitalist Economy*, Oxford: Polity Press.

Graham, E.M. (1975), 'Oligopolistic imitation and European direct investment'. Unpublished Ph.D. thesis, Harvard Graduate School of Business Administration.

Graham, E.M. (1978), 'Transatlantic investment by multinational firms: a rivalistic phenomenon?', *Journal of Post-Keynesian Economics*, vol. 1, no. 1, fall.

Graham, E.M. (1985), 'Intra-industry direct investment, market structure, firm rivalry and technological performance', in Erdilek, A. (ed.), *Multinationals as Mutual Invaders: Intra-Industry Direct Foreign Investment*, London: Croom Helm.

Gray, H.P. (1982), 'Macroeconomic theories of foreign direct investment: an assessment', in Rugman, A.M. (ed.), *New Theories of the Multinational Enterprise*, London: Croom Helm.

Greenaway, D. and Milner, C.R. (1986a), *The Economics of Intra-Industry Trade*, Oxford: Basil Blackwell.

Greenaway, D. and Milner, C.R. (1986b), 'Intra-industry trade: current perspectives and unresolved issues'. Paper presented at the Symposium on Intra-Industry Trade, Abo, Akademi, Finland, April.

Greenaway, D. and Tharakan, P.K.M. (eds, 1986), *Imperfect Competition and International Trade: The Policy Aspects of Intra-Industry Trade*, Brighton: Wheatsheaf.

Grubel, H.G. and Lloyd, P.J. (1975), *Intra-Industry Trade*, London: Macmillan.

Hart, P.E. (1976), 'The dynamics of earnings, 1963–1973', *Economic Journal*, vol. 86, no. 3, September.

Hart, P.E. and Prais, S.J. (1956), 'The analysis of business concentration: a statistical approach', *Journal of the Royal Statistical Society*, series A, vol. 119, no. 2.

Helleiner, G.K. (1981), *Intra-Firm Trade and the Developing Countries*, London: Macmillan.

Helpman, E. and Krugman, P.R. (1985), *Market Structure and Foreign Trade: Increasing Returns, Imperfect Competition, and the International Economy*, Brighton: Wheatsheaf.

Hirsch, S. (1967), *Location of Industry and International Competitiveness*, Oxford: Oxford University Press.

Hirsch, S. (1976), 'An international trade and investment theory of the firm', *Oxford Economic Papers*, vol. 28, no. 2, July.

Hood, N. and Young, S. (1980), *European Development Strategies of US-Owned Manufacturing Companies Located in Scotland*, Edinburgh: HMSO.

Hufbauer, G.C. (1965), *Synthetic Materials and the Theory of International Trade*, London: Duckworth.

Hufbauer, G.C. (1970), 'The impact of national characteristics and technology on the commodity composition of trade in manufactured goods', in Vernon, R. (ed.), *The Technology Factor in International Trade*, New York: Columbia University Press.

Hymer, S. (1975), 'The multinational corporation and the law of uneven development', in Radice, H. (ed.), *International Firms and Modern Imperialism*, Harmondsworth: Penguin.

Hymer, S. (1976), *The International Operations of National Firms: A Study of Direct Investment*, Cambridge, Mass.: MIT Press.

Hymer, S. and Rowthorn, R. (1970), 'Multinational corporations and international oligopoly: the non-American challenge', in Kindleberger, C.P. (ed.), *The International Corporation: A Symposium*, Cambridge, Mass.: MIT Press.

Jenkins, R. (1984), *Transnational Corporations and Industrial Transformation in Latin America*, London: Macmillan.

Jenkins, R. (1987), *Transnational Corporations and Uneven Development: The Internationalisation of Capital and the Third World*, London: Methuen.

Johnson, H.G. (1958), *International Trade and Economic Growth*, London: Allen and Unwin.

Kay, N. (1983), 'Review article: multinational enterprise', *Scottish Journal of Political Economy*, vol. 30, no. 3, November.

Kindleberger, C.P. (1969), *American Business Abroad: Six Lectures on Direct Investment*, New Haven, Conn.: Yale University Press.

Knickerbocker, F.T. (1973), *Oligopolistic Reaction and the Multinational Enterprise*, Boston, Mass.: Harvard University Press.

Knickerbocker, F.T. (1976), 'Market structure and market power consequences of foreign direct investment by multinational companies', Occasional Paper no. 8, Washington, DC: Center for Multinational Studies.

Kojima, K. (1978), *Direct Foreign Investment: A Japanese Model of Multinational Business Operations*, London: Croom Helm.

Kojima, K. and Ozawa, T. (1985), 'Toward a theory of industrial restructuring and dynamic comparative advantage', *Hitotsubashi Journal of Economics*, vol. 26, no. 2, December.

Kregel, J.A. (1977), 'Ricardo, trade and factor mobility', *Economia Internazionale*, vol. 30.

Kroner, M.L. (1980), 'US international transactions in royalties and fees', *Survey of Current Business*, vol. 60, no. 1, January.

Lake, A. (1976a), 'Transnational activity and market entry in the semiconductor industry', National Bureau of Economic Research Working Paper, no. 126, March.

Lake, A. (1976b), 'Foreign competition and the UK pharmaceutical industry', National Bureau of Economic Reseach Working Paper, no. 155, November.

Lall, S. (1976), 'Theories of direct foreign private investment and multinational behaviour', *Economic and Political Weekly*, vol. 11, nos. 31–3, August.

Lall, S. (1978), 'Transnationals, domestic enterprises, and industrial structure in host LDCs: a survey', *Oxford Economic Papers*, vol. 30, no. 2, July.

Lall, S., with Chen, E., Katz, J., Kosacoff, B. and Villela, A. (1983), *The New Multinationals: The Spread of Third World Enterprises*, Chichester: John Wiley and Sons.

Leontief, W.W. (1954), 'Domestic production and foreign trade: the American capital position reexamined', *Economia Internazionale*, vol. 7, no. 1, February.

Linder, S.B. (1961), *An Essay on Trade and Transformation*, New York: John Wiley and Sons.

Magee, S.P. (1977), 'Multinational corporations, the industry technology cycle and development', *Journal of World Trade Law*, vol. 11, no. 4, July–August.

Maneschi, A. (1983), 'Dynamic aspects of Ricardo's international trade theory', *Oxford Economic Papers*, vol. 35, no. 1, March.

Myint, H. (1977), 'Adam Smith's theory of international trade in the perspective of economic development', *Economica*, vol. 44, no. 3, August.

Nelson, R.R. and Winter, S.G. (1977), 'In search of a useful theory of innovation', *Research Policy*, vol. 5, no. 1.

Nelson, R.R. and Winter, S.G. (1982), *An Evolutionary Theory of Economic Change*, Cambridge, Mass.: Harvard University Press.

Newfarmer, R.S. (ed., 1985), *Profits, Progress and Poverty: Case Studies of International Industries In Latin America*, Notre Dame, Ind.: University of Notre Dame Press.

Nicholas, S.J. (1983), 'Agency contracts, institutional modes, and the transition to foreign direct investment by British manufacturing multinationals before 1935', *Journal of Economic History*, vol. 43.

Nicholas, S.J. (1985), 'The theory of multinational enterprise as a transactional mode', in Hertner, P. and Jones, G. (eds), *Multinationals: Theory and History*, Farnborough: Gower.

Nicholas, S.J. (1986), 'Multinationals, transaction costs and choice of institutional form', University of Reading Discussion Papers in International Investment and Business Studies, no. 97, September.

OECD (1979), *Impact of Multinational Enterprises on National Scientific and Technical Capacities: The Food Industry*, Paris: OECD, August.

Ozawa, T. (1974), *Japan's Technological Challenge to the West, 1950-1974: Motivation and Accomplishment*, Cambridge, Mass.: MIT Press.

Ozawa, T. (1979), *Multinationalism, Japanese Style: the political economy of outward dependency*, Princeton: N.J. Princeton University Press.

Ozawa, T. (1982), 'A newer type of foreign investment in Third World resource development', *Rivista Internazionale di Scienze Economiche e Commerciali*, vol. 29, no. 12, December.

Pasinetti, L.L. (1977), *Lectures on the Theory of Production*, London: Macmillan.

Pasinetti, L.L. (1981), *Structural Change and Economic Growth: A Theoretical Essay on the Dynamics of the Wealth of Nations*, Cambridge: Cambridge University Press.

Patel, P. and Pavitt, K. (1986), 'Is Western Europe losing the technological race?', Science Policy Research Unit Discussion Paper, September.

Pavitt, K. (1982), 'R&D, patenting and innovative activities: a statistical exploration', *Research Policy*, vol. 11, no. 1.

Pavitt, K. (1985), 'Patent statistics as indicators of innovative activities: possibilities and problems', *Scientometrics*, vol. 7, nos. 1–2, January.

Pavitt, K. (1987), 'International patterns of technological accumulation', in Hood, N. and Vahne, J.E., (eds), *Strategies in Global Competition*, London: Croom Helm.

Pavitt, K., Robson, M. and Townsend, J. (1987a), 'The size distribution of innovating firms in the UK, 1945–1983', *Journal of Industrial Economics*, vol. 35, no. 3, March.

Pavitt, K., Robson, M. and Townsend, J. (1987b), 'Technological accumulation, diversification and organisation in UK companies, 1945–1983', Science Policy Research Unit Discussion Paper, no. 50, July.

Pavitt, K. and Soete, L.L.G. (1982), 'International differences in economic growth and the international location of innovation', in Giersch, H. (ed.), *Emerging Technologies: Consequences for Economic Growth, Structural Change and Employment*, Tübingen: J.C.B. Möhr.

Pearce, R.D. (1986), 'The internationalisation of research and development by leading enterprises: an empirical study', University of Reading Discussion Paper in International Investment and Business Studies, no. 99, November.

Penrose, E. (1959), *The Theory of the Growth of the Firm*, Oxford: Basil Blackwell.

Phillips, A. (1971), *Technology and Market Structure*, Lexington, Mass.: Lexington Books.

Posner, M.V. (1961), 'Technical change and international trade', *Oxford*

Economic Papers, vol. 13.

Robinson, J. (1974), *Reflections on the Theory of International Trade*, Manchester: Manchester University Press.

Ros, J. (1986), 'Trade, growth and the pattern of specialisation', *Political Economy: Studies in the Surplus Approach*, vol. 2, no. 1.

Rosenberg, N. (1976), *Perspectives on Technology*, Cambridge: Cambridge University Press.

Rosenberg, N. (1982), *Inside the Black Box: Technology and Economics*, Cambridge: Cambridge University Press.

Rosenbluth, G. (1970), 'The relation between foreign control and concentration in Canadian industry', *Canadian Journal of Economics*, vol. 3, no. 1, February.

Rugman, A.M. (1979), *International Diversification and the Multinational Enterprise*, Lexington, Mass.: Lexington Books.

Rugman, A.M. (1980), 'Internalisation as a general theory of foreign direct investment: a reappraisal of the literature', *Weltwirtschaftliches Archiv*, vol. 116, no. 2.

Rugman, A.M. (1981), *Inside the Multinationals: the Economics of Internal Markets*, London: Croom Helm.

Sanna Randaccio, F. (1980), 'European direct investments in US manufacturing'. Unpublished M.Litt. thesis, University of Oxford.

Savary, J. (1984), *French Multinationals*, London: Frances Pinter.

Scherer, F.M. (1983), 'The propensity to patent', *International Journal of Industrial Organisation*, vol. 1, no. 1, December.

Scherer, F.M. (1984), *Innovation and Growth: Schumpeterian Perspectives*, Cambridge, Mass.: MIT Press.

Schumpeter, J.A. (1934), *The Theory of Economic Development*, Cambridge, Mass.: Harvard University Press.

Schumpeter, J.A. (1943), *Capitalism, Socialism and Democracy*, New York: Harper and Row.

Servan-Schreiber, J.-J. (1967), *Le Défi Américain*, Paris: Editions Denoel.

Soete, L.L.G. (1980), 'The impact of technological innovation on international trade patterns: the evidence reconsidered'. Paper presented at the OECD Science and Technology Indicators Conference, Paris, September.

Soete, L.L.G. (1981), 'A general test of technology gap trade theory', *Weltwirtschaftliches Archiv*, vol. 117, no. 4.

Soete, L.L.G. and Wyatt, S.M.E. (1983), 'The use of foreign patenting as an internationally comparable science and technology output indicator', *Scientometrics*, vol. 5, no. 2.

Steindl, J. (1952), *Maturity and Stagnation in American Capitalism*, Oxford: Oxford University Press.

Stiglitz, J.E. (1987), 'Learning to learn, localised learning and technological progress', in Dasgupta, P. and Stoneman, P. (eds), *Economic Policy and Technological Performance*, Cambridge: Cambridge University Press.

Stopford, J.M. and Turner, L. (1986), *Britain and the Multinationals*, Chichester: John Wiley and Sons.

Sugden, R. (1983), 'Why transnational corporations?', *Warwick Economic Research Paper*, no. 222.

Sutcliffe, C.M.S. and Sinclair, M.T. (1980), 'The measurement of seasonality within the tourist industry: an application to tourist arrivals in Spain', *Applied Economics*, vol. 12, no. 4, December.

Sylos-Labini, P. (1984), *The Forces of Economic Growth and Decline*, Cambridge, Mass.: MIT Press.

Sylos-Labini, P. (1989), 'Oligopoly and technical progress: a critical reconsideration after thirty years', in Arcangeli, F., David, P.A. and Dosi, G. (eds), *Modern Patterns in Introducing and Adopting Innovations*, Oxford: Oxford University Press.

Teece, D.J. (1977), 'Technology transfer by multinational firms: the resource costs of transferring technological know-how', *Economic Journal*, vol. 87, no. 2, June.

Teece, D.J. (1983), 'Technological and organisational factors in the theory of the multinational enterprise', in Casson, M.C. (ed.), *The Growth of International Business*, London: Allen and Unwin.

Teece, D.J. (1989), 'Capturing value from technological innovation: integration, strategic partnering and licensing decisions', in Arcangeli, F., David, P.A. and Dosi, G. (eds), *Modern Patterns in Introducing and Adopting Innovations*, Oxford: Oxford University Press.

Thweatt, W.O. (1976), 'James Mill and the early development of comparative advantage', *History of Political Economy*, vol. 8, no. 2, Summer.

Usher, A.P. (1929), *A History of Mechanical Inventions*, Cambridge, Mass.: Harvard University Press.

Vernon, R. (1966), 'International investment and international trade in the product cycle', *Quarterly Journal of Economics*, vol. 80, no. 2, May.

Vernon, R. (1971), *Sovereignty at Bay*, Harmondsworth: Penguin Books.

Vernon, R. (1974), 'The location of economic activity', in Dunning, J.H. (ed.), *Economic Analysis and the Multinational Enterprise*, London: Allen and Unwin.

Vernon, R. (1979), 'The product cycle hypothesis in the new international environment', *Oxford Bulletin of Economics and Statistics*, vol. 41, no. 4, November.

Walker, W. (1979), *Industrial Innovation and International Trading Performance*, Greenwich, Conn.: JAI Press.

Walsh, V.C. (1979), 'Ricardian foreign trade theory in the light of the classical revival', *Eastern Economic Journal*, vol. 5, no. 3.

Warren, B. (1980), *Imperialism: Pioneer of Capitalism*, London: Verso.

Wells, L.T. (ed., 1972), *The Product Life Cycle and International Trade*, Boston, Mass.: Harvard Univesity Press.

Williamson, O.E. (1975), *Markets and Hierarchies: Analysis and Antitrust Implications*, New York: Free Press.

Wyatt, S.M.E. with Bertin, G. and Pavitt, K. (1985), 'Patents and multinational corporations: results from questionnaires', *World Patent Information*, vol. 7, no. 3.

Index

absolute advantage, 180, 182
Acocella, N., 98, 128
agglomeration, 6, 13, 139, 161
Aliber, R. Z., 78, 189, 203, 206, 211
American Challenge, 1, 2, 51, 58, 71, 74, 91, 137
American firms *see* USA
anti-competitive effects *see* market power
appropriability, 10, 190–1
Archibugi, D., 124, 128
Arthur, W. B., 17, 29

Bain, J. S., 197, 217
Balassa, B., 19, 121
Belgian firms *see* Belgium
Belgium, 23, 24, 27, 32, 36, 37, 38, 40, 44, 78, 84–6
bivariate distributions *see* Galtonian regression
Britain *see* UK
British firms *see* UK
Buckley, P. J., 10, 11, 55, 60, 119, 189, 193, 194, 195, 208, 213, 214

Canada, 23, 26, 31, 32, 33, 35, 36, 37, 38, 39, 40, 41
Canadian firms *see* Canada
Cantwell, J. A., 4, 13, 54, 87, 92, 94, 97, 98, 99, 120, 139, 158, 161, 162, 188, 189, 203

capital accumulation, 7, 8, 9, 14, 106, 120, 141, 161, 163, 164, 166, 167, 176, 180, 207
Casson, M. C., 10, 11, 54, 55, 119, 189, 193, 194, 205, 208, 213, 214, 215, 216
catching up: of countries, 49, 50, 63, 74, 75–7, 111, 115, 144–5; of firms, 5, 50, 78–91, 111, 115
Caves, R. E., 8, 10, 15, 119, 148, 189, 193
centres of innovation, 3, 6, 10, 11, 13, 18, 138–9, 154, 157, 161, 162, 177–9, 181, 207; *see also* location of technological activity
classical economists, 7, 52–3, 162, 176, 192, 214
Clegg, L. J., 89, 96
Coase, R. H., 193, 216
Coasian theory of the firm *see* internalisation
collusion *see* market power
comparative advantage: of countries, 19, 121, 124–7, 180–1, 182, 201, 202; of firms, 17, 50, 62–3, 68, 121, 125, 127–33; *see also* RTA index
competition, 3, 4, 12, 14, 192, 196, 217; *see also* oligopolistic competition; technological competition

competitiveness: of countries,
4–5, 75–84, 91, 93, 97,
111–15, 124–7, 161–2, 179,
181, 183, 184–5, 218; of firms,
2, 5, 73, 74, 78–88, 90–2, 93,
97, 102–17, 120, 127–36,
140–1, 178–9, 218; *see also*
growth; market shares
concentration *see* industrial
concentration
Corley, T. A. B., 188
Cowling, K., 12, 95, 189, 194,
196, 197, 204, 218
cross-investments *see*
intra-industry production
cross-licensing *see* technological
cooperation; *see also* licensing
cumulative causation, 6, 63, 73,
87–8, 139, 140, 161, 172–85
cumulativeness in innovation, 4,
7, 11, 16–17, 18–19, 25, 29,
30, 33, 39, 42, 43, 45, 47, 50,
51, 139–41, 158, 160, 177

differentiated technological
change: of countries, 18,
137–8, 139, 140, 148, 158,
161, 177–9; of firms, 8, 10–11,
16, 18, 137, 139–40, 177–9,
216
Dosi, G., 11, 15, 16
Dunning, J. H., 13, 15, 72, 73,
78, 86, 87, 92, 97, 98, 99,
104, 119, 120, 139, 143, 150,
153, 158, 159, 188, 189, 190,
202, 203, 218
Dutch firms *see* Netherlands

eclectic paradigm, 72, 188,
189–93, 198–9, 206, 211
economic growth *see* growth
economies of scale, 8–9, 13, 54,
59, 86, 95, 110
economies of scope, 13, 95, 141,
190, 193, 209

employment, 8, 164, 165, 174–6,
183, 194, 197–8; and wages,
95, 166, 175–6, 183, 194,
197–8
entrepreneurship, 12, 57, 60, 214
Europe, 1, 5, 22, 49, 50, 51, 58,
62, 68, 71, 75–92, 93, 111,
145–7, 149–50, 152–3, 158,
180, 198, 199, 200, 202, 204,
215–16
European Economic Community,
1, 75, 77, 78, 88, 127
evolutionary approach, 3,
119–20, 186–8, 206–7, 214,
218
export-platform investment,
200–1, 202, 210
exports *see* international trade

firm-specific technology *see*
differentiated technological
change
foreign-owned firms *see* inward
international production
foreign patenting, 19, 21, 32–3,
122, 129
foreign technologies, 3, 13, 19,
46, 57, 139, 151
France, 23, 26, 32–3, 35, 36, 40,
41, 70–1, 75, 84–6, 94, 101,
102, 103, 105, 109, 110, 111,
112, 115, 116, 117, 123, 127,
128–9, 134, 144, 145, 146,
147, 149
Franko, L. G., 54, 55, 60
Freeman, C., 12, 15, 57, 95, 183
French firms *see* France

Galtonian regression, 4, 25,
28–31, 108–9; *see also* normal
distribution
German firms *see* Germany
(West)
Germany (West), 14, 17, 19, 23,
24, 26, 32, 35, 36, 37, 38, 40,

41, 42, 43, 70–1, 75, 84–6, 88, 90, 93, 101, 102, 103, 105, 108, 109, 110, 111, 112, 115, 116, 117, 122–3, 125, 126, 127, 128, 129, 131, 132, 133, 134, 135, 136, 144, 145, 146, 147, 153–7

global competition *see* international industries; *see also* nationalised investment

Goodwin, R. M., 162

Graham, E. M., 54, 141, 189, 203, 204

growth: of countries, 6, 12, 14, 49–50, 52, 65–7, 75–7, 94, 161, 162, 169–70, 171–6, 182; of firms, 8, 65–7, 74, 93–6, 119, 120, 134–5, 177–80, 191–9, 203–6, 207–12, 214–17; *see also* competitiveness

growth cycle, 162, 166, 173

Grubel, H. G., 142, 143

Hart, P. E., 4, 25, 28

Helpman, E., 54, 159

Hirsch, S., 72, 119, 189, 201

history of innovation, 19, 21–4, 31–7, 60–1

host country impacts *see* competitiveness

Hufbauer, G. C., 72

Hymer, S., 2, 74, 188, 189, 195, 197, 199, 204, 211, 213, 217, 218

Hymer's theory of the firm *see* market power

import-substituting investment, 54, 55, 118, 200, 201, 202

incremental technological change, 16, 17–18, 25, 29–30, 33, 37, 39, 42, 45, 47

indigenous firms, 2, 64, 66, 71, 73, 74, 79–88, 91, 154, 178

industrial concentration, 2, 3, 58, 196, 197

industrial pattern of innovation, 4, 16–17, 18, 19, 24, 25, 26–7, 28–9, 31–45, 50, 110, 120, 123–5, 137, 160, 181

industrial structure, 4, 23, 43, 102–3

information, 11

internalisation, 10–11, 55, 56, 119, 132, 190–2, 193–5, 198, 203, 206, 208, 209, 210, 212–13, 214–16

international centres of innovation *see* centres of innovation

international competition *see* competition

international competitiveness *see* competitiveness

international industries, 3, 13, 59, 61, 62, 94–6, 120, 141, 146, 148, 153, 157, 160–1, 173–80, 181, 197, 203–6

international production data, 5, 96, 97–101

international trade, 5, 55, 56, 64–7, 75–86, 97, 111–15, 116, 124–7, 130–3, 161, 167–76; *see also* theories of international trade

intra-industry production, 6, 51, 54, 59, 63, 67, 68, 120, 137–59, 161, 203, 204

intra-industry trade, 54, 63, 120, 138, 146–53

inward international production, 5, 51, 77–88, 143, 145, 160, 178

Italian firms *see* Italy

Italy, 23, 27, 32, 36, 38, 40, 41, 44, 70–1, 75, 78, 84–6, 93, 101, 102, 103, 105, 108, 109,

110, 111, 112, 115, 116, 117, 123–4, 127, 128, 134, 144, 145, 146, 147, 149

Japan, 1, 2, 14, 17, 23, 24, 27, 34, 38, 40, 41, 42, 43, 44, 46, 49, 50, 58, 93–4, 101, 102, 103, 105–6, 109, 110, 111, 112, 115, 116, 117, 124, 126, 127, 128, 129, 131, 132, 133, 134, 135, 136, 145, 146, 147, 180, 198, 200, 201, 202
Japanese firms *see* Japan
Jenkins, R., 189, 192, 198, 205, 219

Kindleberger, C. P., 2, 197, 211
Knickerbocker, F. T., 2, 15, 54, 189
Kojima, K., 189, 200, 201, 202, 210, 213
Krugman, P. R., 54, 159

labour productivity *see* productivity growth
Lake, A., 74, 86, 87
Lall, S., 15, 94, 158, 195
licensing, 62, 66–7, 88–91, 96, 119, 191, 202, 217
limit cycle, 165
Lloyd, P. J., 142, 143
location advantages, 55–6, 62, 77–8, 101, 116–17, 120, 132, 154, 190, 206, 207, 212–13
location of technological activity, 3, 6, 13, 18, 86–8, 139, 140, 151, 153, 156–9, 162, 177–80, 184, 191, 209, 218; *see also* centres of innovation

machinery, 7
macroeconomic theories *see* theories of international production

manufacturing as source of innnovation, 7, 151–3
market power, 2, 3, 12, 14, 189, 192, 195–9, 203, 204, 206, 211, 213, 217
market shares, 5, 60, 64, 65, 94, 102–10, 115, 127–33, 160, 178, 204, 207; *see also* competitiveness
market structure, 2, 6, 56, 58, 192, 197, 211, 216; *see also* industrial concentration
Marx, K., 7, 8
MNC international networks, 3, 4, 5, 10, 11, 12, 14, 58, 69, 93, 95, 114–15, 127, 135, 141, 177–80, 187, 194, 200, 209, 211
mobility effect, 31, 33, 37, 39, 42, 43, 44, 111
monopoly power *see* market power

Nelson, R. R., 8, 214
Netherlands, 23, 27, 34, 38, 40, 41, 75, 78, 84–6
Newfarmer, R. S., 15, 189, 196, 197, 213
Nicholas, S. J., 214, 216
non-affiliate licensing *see* licensing
normal distribution, 25, 28, 31–2, 34, 35–6, 38
Norman, G., 143, 150, 159

oligopolistic competition, 1, 12, 54, 59, 94, 141, 146, 189, 198, 203–5, 217; *see also* competition; international industries
organisational change, 3, 9, 94, 184
outward international production, 67, 91, 137, 143, 145, 154, 178–9

ownership advantages, 55, 56, 60–1, 62, 64, 65, 119, 132, 190–3, 206–12; *see also* technological advantage
Ozawa, T., 189, 200, 201, 202, 210, 213

Pasinetti, L. L., 53, 182, 202
Patel, P., 2, 45
patenting, 4, 13, 20–4, 32–3, 37, 38, 46, 50, 122, 128, 207; *see also* foreign patenting; propensity to patent; RTA index
path-dependent processes, 17, 29
Pavitt, K., 2, 7, 16, 20, 21, 30, 37, 38–9, 45, 47, 130, 134, 139
Pearce, R. D., 98, 99, 104, 159
Penrose, E., 204, 216
Perez, C., 12, 95, 183
Prais, S. J., 4, 25
product cycle model, 51–61, 68, 118, 189, 199, 200, 201, 202
production engineering, 3, 19, 46, 134
productivity growth, 8, 10, 43, 162, 166, 167, 172, 180–1, 182, 184
profits, 8, 59, 95, 162, 163, 164, 184, 194, 196, 197–8, 204, 207, 217; *see also* share of wages
propensity to patent, 20–1
Punzo, L. F., 162

rate of innovation, 13, 17, 43–4, 46, 50, 134–5, 167
rationalised investment, 13, 78, 79, 135
regression effect, 28–9, 33, 36, 37, 39, 40
research and development (R&D), 9, 13, 18–19, 21, 46, 57, 135,

139, 151–2, 159
research-intensive production, 1, 9, 68, 139, 161, 177–9, 210
reverse technology transfer, 141, 151, 158
Ricardo, D., 7, 52, 162, 180, 201
Robson, M., 130, 134
Ros, J., 182
Rosenberg, N., 7, 8, 16, 17, 212
Rowthorn, R., 74, 204
RTA index, 19–20, 22–44, 65, 120–33, 156–7
Rugman, A. M., 189, 194, 211

Sanna Randaccio, F., 60, 141, 204
Savary, J., 98, 196
scale economies *see* economies of scale
Scherer, F. M., 7, 20, 151
Schumpeter, J. A., 12, 57, 58
Schumpeterian approach, 12, 49–51, 57, 60, 94, 134–5
sectoral distribution of innovation *see* industrial pattern of innovation
Servan-Schreiber, J.-J., 1
share of wages, 163, 164, 165, 166, 169, 171, 172, 173, 183
Smith, A., 7, 162, 197
Soete, L. L. G., 4, 19, 20, 30, 118, 183
Sugden, R., 12, 189, 194, 196, 197, 204, 218
surplus *see* profits
Sylos-Labini, P., 12

technological accumulation, 7–14, 16–19, 29–30, 42–3, 45–7, 50, 61–3, 74, 86, 103, 106, 120, 140, 141, 158, 160, 161, 162, 163, 164, 165, 166, 167–80, 181, 184, 186–7, 189, 205, 206, 207, 208, 211–12,

214, 216, 217, 218; *see also*
cumulativeness in innovation;
differentiated technological
change; incremental
technological change
technological advantage, 4, 5, 10,
16, 18–19, 21, 25–9, 31–45,
50, 51, 62–3, 68, 86–7, 91,
103, 105–6, 111, 118, 120,
122–36, 138, 153–4, 156–7,
191–2, 208, 209, 211; *see also*
ownership advantages; RTA
index
technological competition, 2, 3,
4, 13, 14, 49, 51, 52, 58,
61–8, 74, 94–5, 109, 120, 135,
136, 138, 148, 151, 156–7,
158, 160–1, 198, 200, 203,
207, 217, 218; *see also*
competition
technological competitiveness *see*
competitiveness
technological cooperation, 7, 13,
45, 120, 135, 140, 212, 217
technological diversification *see*
technological specialisation
technological gap, 1, 73, 74, 118,
200
technological interrelatedness, 13,
17, 25, 47, 140
technological leadership, 15, 49,
50, 58, 60, 61, 62, 65, 200
technological paradigm, 12, 15,
19, 46, 95
technological specialisation, 3,
12, 13, 17, 24, 30–1, 33–4, 37,
39, 42, 46–7, 110, 122, 182
technology diffusion, 9, 157, 158
technology transfer, 9, 10–11,
45–6, 158; *see also* reverse
technology transfer
Teece, D. J., 10, 189, 214, 216,
218

theories of the firm *see*
internalisation *and* market
power
theories of industrial
organisation, 2, 188, 192,
199–203, 206, 213
theories of innovation, 11–12,
53; *see also* product cycle
model; tehnological
accumulation
theories of international
production, 118–19, 186–218;
macroeconomic theories, 187,
189, 199–203, 206, 213; *see
also* eclectic paradigm;
internalisation; market power;
product cycle model;
technological accumulation
theories of international trade,
51–2, 53, 118–19, 162, 180–2,
187, 199, 200; *see also* product
cycle model
Thweatt W. O., 52, 162
Tolentino, P. E. E., 94, 189
Townsend, J., 130, 134
trade *see* international trade
trade imbalances, 142, 146,
170–6, 181, 184, 199, 200,
202
trade unions *see* employment

UK, 14, 23, 24, 26, 32, 35, 36,
37, 38, 39, 40, 41, 42–3, 44,
70–1, 84–6, 87–8, 90, 101,
102, 103, 105, 108, 110, 111,
114, 115, 116, 123, 127, 128,
129, 132, 133, 134, 135, 144,
145, 146, 147
UK firms *see* UK
USA, 1, 5, 14, 19, 20, 21, 23, 27,
38, 39, 40, 41, 42, 49, 50, 51,
58, 62, 69–71, 75–92, 93, 101,
102, 103, 105, 106, 108, 109,

110, 111, 114, 115, 116, 122, 125, 129, 132, 133, 134 ,135, 143, 145, 146, 147, 149, 151, 153–7, 158, 180, 198, 199, 200, 202, 204, 215–16
US firms *see* USA
Usher, A. P., 8

Vernon, R., 52, 55, 58, 72, 189, 199, 201, 202, 203
vertical integration, 149, 193, 216
vertical linkages, 151, 153
vicious circle *see* cumulative causation

virtuous circle *see* cumulative causation

wages *see* employment; *see also* share of wages
Walker, W., 79
Wells, L. T., 72
West Germany *see* Germany (West)
Williamson, O. E., 10, 189, 194, 214
Winter, S. G., 8, 214
Wyatt, S. M. E., 20, 136